Key Questions in Environmental Toxicology
A Study and Revision Guide

Key Questions in Environmental Toxicology

A Study and Revision Guide

J.P.F. D'Mello
Formerly of SAC, University of Edinburgh King's Buildings Campus, Edinburgh, UK

CABI is a trading name of CAB International

CABI
Nosworthy Way
Wallingford
Oxfordshire OX10 8DE
UK

Tel: +44 (0)1491 832111
E-mail: info@cabi.org
Website: www.cabi.org

CABI
WeWork
One Lincoln St
24th Floor
Boston, MA 02111
USA

Tel: +1 (617)682-9015
E-mail: cabi-nao@cabi.org

A catalogue record for this book is available from the British Library, London, UK.

References to Internet websites (URLs) were accurate at the time of writing.

ISBN-13: 9781789248524 (paperback)
         9781789248531 (ePDF)
         9781789248548 (ePub)

DOI: 10.1079/9781789248548.0000

Commissioning Editor: Alexandra Lainsbury
Editorial Assistant: Lauren Davies
Production Editor: James Bishop

Typeset by SPi/Straive Pondicherry, India
Printed and bound in the UK by CPI Group (UK) Ltd, Croydon, CR0 4YY

Contents

Preface

Introduction

This edition of *Key Questions in Environmental Toxicology* is designed as a revision aid for students in the first 2 years of undergraduate degree courses in environmental science, medical toxicology, applied chemistry and ecology. Other potential users might include geography and geology graduates wishing to transfer to taught Masters courses in the aforementioned disciplines. This book should be used in conjunction with the CABI text, *Introduction to Environmental Toxicology* (D'Mello, 2020a) which is the template for *Key Questions in Environmental Toxicology*. It is perceived that both titles should represent user-friendly editions, enabling students with appropriate university-entrance qualifications to integrate fundamental concepts within modules of existing BSc/BS degree courses. Those likely to benefit most are students with a sound knowledge of organic chemistry and biology to at least A level/Higher standards in the UK and equivalent secondary-school attainments in other countries.

Learning Objectives

The primary objectives in this and in the companion volume *Introduction to Environmental Toxicology* (D'Mello, 2020a) are to provide university and college students with a general overview and understanding of:

- key issues in environmental toxicology;
- the actual or potential harm caused by pollution on human health;
- the existential risks for wildlife arising from anthropogenic contaminants in soil, freshwater and marine ecosystems; and
- knowledge of specific case studies and major contamination incidents.

Additional objectives are to develop skills in comprehension, competence and communication in the analysis of pollution events, historical and contemporary. These objectives will be fulfilled and enhanced if degree courses are underpinned by tutorials, seminars and imaginative practical sessions in the laboratory. Participation in field studies and visits to local environmental protection agencies would provide valuable information to augment the learning experience.

Learning Outcomes

Successful completion of assessments set out in *Key Questions in Environmental Toxicology*, together with attendance at seminars and tutorials as well as submission of essays and laboratory reports should enable students to:

- emphasize the three environmental pollution emergencies that have been outlined in my *Introduction* book;
- critically appraise the limited approach of international agencies and high-profile pressure groups in highlighting climate change as the sole manifestation of the impending environmental crisis;
- challenge the current risk-management agenda that is almost entirely dictated by efforts to curb the emissions of detrimental greenhouse gases;
- argue that there is a critical need also to consider the anthropogenic production of a diverse range of other pollutants now unequivocally associated with human health disorders and ecotoxicity;
- identify and classify pollutants according to recognized systems to distinguish between legacy, gaseous, organic, inorganic, persistent and radioactive pollutants;
- question the prevailing consensus that the 'carbon footprint' and the 'carbon-neutral' approaches should focus entirely on monitoring and limiting tropospheric concentrations of carbon dioxide and methane;
- demonstrate that existing scientific evidence indicates that surveillance and regulations should be extended to a wide range of other contaminants in order to reduce pollution in diverse ecosystems;
- contend that a significant proportion of complex pollutants contain carbon and thus the carbon trail should be viewed in a more comprehensive perspective in order to mitigate risks for human health and to enhance survival of endangered wildlife species;
- discuss that in addition, emissions of nitrogen dioxide, sulfur dioxide, ozone, particulates, heavy metals and radiation are definitively associated with morbidity in humans; and
- use case studies and contemporary contamination incidents to support the arguments advanced in the foregoing learning outcomes.

Performance Criteria and Progression

Assessments of the above learning outcomes should, ideally, be conducted by submission of appropriate practical reports and, in addition, extended response questions set in formal examination papers. This was the approach I used in my degree course in *Environmental Protection and Management* delivered under the regulations of the University of Edinburgh.

Satisfactory achievement in these learning outcomes should also enable students to progress to the final year of the different courses cited above or transfer to the first year of taught MSc/MS degree courses. However, the latter option might require knowledge at a more advanced level as presented in *A Handbook of Environmental Toxicology* (D'Mello, 2020b) which I edited for CABI in 2020.

How to Use *Key Questions in Environmental Toxicology*

This edition is set out in the standard format adopted in a wide range of books on revision questions in different scientific disciplines. Five types of questions are used in this volume including:

- fill-in-the-gap;
- multiple choice;
- short answer;
- essay; and
- case study analysis.

In the multiple-choice questions, four options are presented and students should select all that apply as there may be more than one correct answer. Students should attempt questions in the sequence presented and avoid glancing at subsequent questions and choices as these may affect selection of correct answers. In order to enhance the revision process, students are encouraged to justify answers for the fill-in-the-gap and multiple-choice questions.

Students may wish to compare their answers with the detailed solutions which appear at the end of each chapter. Due to regular updating of emerging data in a fast-moving field, answers provided here may differ in detail from material published in the companion book *Introduction to Environmental Toxicology*. Essay questions and case study assignments, particularly those in Chapter 10, are designed to be more challenging by examining skills in overall integration and interpretation of emerging scientific evidence rather than recall of facts presented in preceding chapters.

In addition, supplementary questions appear throughout this volume at key points in the text, figures and tables. These are intended to encourage students to read around the subject and to provide topics for tutorials, seminars and revision sessions. Answers to selected supplementary questions are provided in the Appendix.

Environmental Diary

In order to establish relevance to everyday issues, students are strongly advised to record emerging incidents and announcements in an environmental diary. Items should be selected on the basis of supporting evidence published in scientific journals. It is clear now, more than ever before, that environmental toxicology cannot be perceived purely as an academic discipline. Information recorded in the diary should be used to supplement evidence in the essay questions set out in this volume and in university examination papers. The author's diary appears in Chapter 10, this volume, with appropriate commentary relating to toxicological implications of unfolding issues.

References

D'Mello, J.P.F. (2020a) *Introduction to Environmental Toxicology*. CAB International, Wallingford, UK.

D'Mello, J.P.F. (ed.) (2020b) *A Handbook of Environmental Toxicology: Human Disorders and Ecotoxicology*. CAB International, Wallingford, UK.

Disclaimer

This book necessarily contains references to commercial products and contamination incidents linked to multinational conglomerates. It is emphasized that all examples used in this volume are based on evidence already in the public domain, with commentaries from respected sources including research journals, prominent medical scientists and ecologists. This material is presented entirely for educational reasons and not to embarrass individuals or companies deemed by others to be at fault.

Acknowledgement

I wish to record my sincere gratitude for the expert guidance and encouragement provided by Alex Lainsbury, the commissioning editor at CABI.

1 Basic Principles

Questions

1.1 Overview

1.1.1. The history of global environmental pollution has been shaped by significant events including: the 'Great Smog of _____' in 1952; the 1986 Chernobyl nuclear plant explosion in _____; the *Exxon Valdez* oil spill, off the coast of _____ in 1989; the _____ _____ explosion in the Gulf of Mexico in 2010.

1.1.2. The current environmental debate, focusing entirely on the failure to reduce global greenhouse gas emissions, has overshadowed the urgent need to curb exposures of humans and wildlife to other _____. Additional human health disorders are often associated with exposures to _____ and _____ _____.

1.2 Methodology

1.2.1. The existing protocols used to characterize the toxic risks of biogenic contaminants and environmental pollutants are:

  (a) absolutely reliable

  (b) totally inadequate

  (c) subject to continuous evaluation

  (d) only relevant for greenhouse gases

© J.P.F. D'Mello 2022. *Key Questions in Environmental Toxicology* (J.P.F. D'Mello)
DOI: 10.1079/9781789248548.0001

1.2.2. The term LD_{50} signifies:

    (a) least difference between 50 species of test organisms

    (b) lethal dose to kill 50% of a population of test organisms

    (c) local dose to harm 50% of the population

    (d) lethal defects in 50 individuals in a population exposed to polluted air

1.2.3. The term LC_{50} signifies:

    (a) lethal concentration causing mortality in 50% of a population of organisms

    (b) least concentration causing mortality in 50% of a population of organisms

    (c) lead concentrations exceeding 50 ng l^{-1} blood

    (d) liquid chromatography of 50 ambient pollutants

1.2.4. Toxicity of chemical and physical stressors resulting in lethal end points is:

    (a) exclusively an occupational hazard

    (b) restricted entirely to laboratory investigations

    (c) a regular occurrence with diverse implications

    (d) now consigned to history

1.2.5. Lethality is still used as a fundamental criterion in toxicology. Discuss this statement, commenting also on ecological implications of current protocols.

1.2.6. Write a brief note on how chronic toxicity of contaminants and pollutants is generally assessed.

1.3 Exposure Pathways

1.3.1. In humans and other vertebrates, exposures to contaminants and pollutants may occur via:

    (a) ingestion of food

    (b) water intake

    (c) contact with other individuals

    (d) inhalation

1.4 Interactions

1.4.1. In additive interactions between two toxic pollutants, the resulting adverse effects on populations of organisms are:

(a) the sum of the individual effects

(b) greater than the sum of the individual effects

(c) less than the sum of individual effects

(d) of minor significance in risk assessment

1.4.2. In synergistic interactions between two toxic pollutants, the resulting effects on populations of organisms are:

(a) the sum of the individual effects

(b) greater than the sum of the individual effects

(c) less than the sum of individual effects

(d) of no consequence in risk assessment

1.4.3. Comment on the potential or actual significance of interactions in determining health outcomes for humans exposed to toxic pollutants.

1.5 Cellular and Metabolic Responses

1.5.1. Write a short note on the cellular and metabolic responses that might occur on exposure of living organisms to toxic pollutants.

1.6 Environmental Fate

1.6.1. Explain why knowledge of the ultimate fate of pollutants is of critical importance in environmental protection and management.

1.7 Risk Assessment and Regulation

1.7.1. Write a short note on biomonitoring as an aid in risk assessment.

1.7.2. Outline the regulatory networks currently in existence to control risks linked to toxic pollutants.

Answers

1.1 Overview

1.1.1. Insert, respectively: London; Ukraine; Alaska; *Deepwater Horizon*. These events raised awareness of:

- emissions of toxic gases and particulates;
- ionizing radiation exposures;
- crude oil pollution; and
- ecological impacts.

1.1.2. Insert: pollutants; radon; ultraviolet radiation.

The disproportionate emphasis on climate change, relative to environmental toxicology can no longer be justified in the light of unequivocal evidence of profound adverse effects caused by:

- nitrogen dioxide;
- sulfur dioxide;
- persistent organic compounds (POPs);
- fossil fuel pollutants;
- toxic metals;
- consumer products and lifestyle choices; and
- radiation in its three forms.

1.2 Methodology

1.2.1. Select (c).

The protocols used to evaluate the toxic characteristics of contaminants and pollutants are under regular review. Traditional methods have relied on assessments of hazards associated with acute and chronic exposures, as, for example, in the safety evaluation of pharmaceuticals, food additives and pesticides.

In environmental toxicology, however, risk assessment methodologies are required to accommodate a wide spectrum of end points relevant to specific environmental stressors.

Appropriate end points might include one or more of the following:

- dermal lesions;
- ocular dysfunction;
- immunosuppression;
- reproductive disruption;
- carcinogenesis;

- respiratory disorders;
- cardiovascular disease;
- neurodegenerative disorders; and
- mortality.

In ecological settings, relevant end points generally include:

- lethality;
- reproductive disruption;
- relative abundance of endangered wildlife species;
- behavioural responses;
- recovery of wildlife species following significant pollution events;
- bleaching of coral species; and
- habitat degradation and loss.

Acute and chronic assays can only provide limited data in respect of the end points cited above and the development of more sophisticated techniques is, therefore, required. Furthermore, extrapolation of data obtained with laboratory models to humans is problematic. Thus, unravelling interactions between ambient pollutants and underlying conditions such as respiratory or cardiovascular disease is more challenging in standard toxicity assays.

Another major limitation of traditional methodologies has been reliance on the use of rodent models in laboratory tests. Such assays are not only expensive to conduct but also provoke protests from certain sections of society. Currently, there is considerable interest in the use of a zebra fish model to replace rodents in laboratory assays. Reference to this model will appear in subsequent chapters in this volume.

It is often suggested that greater reliance should be placed on *in vitro* investigations using, for example, human hepatoma cells (Arumugam *et al.*, 2021). However, inter-organ dynamics and biotransformation cannot be fully replicated under these conditions.

Increasingly, researchers are turning to epidemiological models, as exemplified in previous well-known correlations linking tobacco smoking with lung cancer as well as excessive alcohol consumption with liver cirrhosis.

- Can you define 'epidemiology'?

1.2.2. Select (b).

LD_{50} represents the lethal dose to kill 50% of a population of test organisms. The LD_{50} assay is a quantitative assessment of acute toxicity of a specific/suspected stressor.

Lethality data are generally obtained with a population of test organisms in a dose–response experiment. Graded doses of the toxicant under investigation are used to determine the effects on potential mortality for the test organisms. The dose–response curve so obtained will correspond to a sigmoid shape, with low doses causing small but perceptible effects, followed by a steep increase in lethality up to an asymptote of 100% mortality in the population of organisms.

Despite well-established limitations, LD_{50} assays are still used in research trials, particularly for emerging toxins and pesticides. It is worthwhile reading the paper by Boente-Juncal et al. (2020) for derivation of no observed adverse effect level (NOAEL) from the LD_{50} data.

- Can you draw a sigmoid curve?
- Has the application of NOAEL been explained by your course lecturer?

1.2.3.   Select (a).

In an ecological context, acute toxicity results are reported as LC_{50} values and relate to the concentration of the test compound required to kill 50% of the population in aquatic or atmospheric systems. Favoured test organisms used in environmental biomonitoring include species of *Daphnia*. The paper by Kergaraval et al. (2021) provides an update on the use of this species in the ecotoxicity of antibiotics.

- Can you name another species used in biomonitoring?

1.2.4.   Select (c).

Toxicity of chemical and physical stressors resulting in lethality is a regular occurrence, impacting human communities and wildlife. Indeed, toxicology is founded on numerous historical, and occasionally anecdotal, examples of harm. There are still regular occurrences that raise issues of concern and requiring further investigation, and not just for humans. Examples below are selected to illustrate diverse impacts of:

- vehicular pollution:
  - gaseous;
  - particulates;
- radionuclide emissions;
- climate change and algal blooms:
  - toxins;
  - anoxic effects; and
- pesticides.

The Great Smog of London claimed the lives of approximately 4000 individuals in 1952, attributed to the use of coal for domestic heating and industrial purposes. Fatalities continue to be registered in London and across the globe due to road traffic pollution produced during gasoline combustion in vehicle engines. Premature mortality is a particular feature for individuals with underlying conditions such as respiratory or cardiovascular disease residing in polluted cities.

In 2020, alerts were issued over predictions of an 'ecological disaster' in Kenyan Rift Valley lakes relating to deaths of over 300 elephants at waterholes in Central Africa. It is now emerging that the alkaline stagnant waters in these lakes promotes the proliferation and growth of toxic algal blooms.

In other ecological settings, there are serious concerns over the virtual extinction of several species of insects, attributed to increasing and indiscriminate use of pesticides.

- What are the major classes of pesticides?

1.2.5. Although much maligned by toxicologists, particularly regarding acute tests, lethality is still a fundamental criterion used to determine the efficacy of pesticides. Indeed, the US Environmental Protection Agency (EPA), as well as other regulatory authorities, regularly evaluate pesticides in the light of lethality data.

Relevant issues to consider include:

- duration of assays;
- impact on non-target species;
- a case study:
  - neonicotinoid insecticide; and
- limitations.

Toxicity evaluations by manufacturers typically focus on relatively short-term tests with mammalian models (rodents) to establish risks for humans consuming pesticide-treated foods. However, the impact on invertebrates, aquatic animals and other non-target species is often not considered in satisfactory detail. The observations in a case study illustrate the point regarding a widely used neonicotinoid insecticide introduced into Japan in 1993. Numbers of arthropods in Lake Shinji were decimated almost immediately leading to a collapse of the food web in the aquatic ecosystem, thereby adversely impacting on populations of fish. These effects were attributed to leaching of the pesticide into the lake. Neonicotinoids and other pesticides are fatal for bees and butterflies, while other evidence implies detrimental effects on bird populations.

More significantly, however, lethality tests cannot be used to determine effects of pesticides as carcinogens or as agents promoting or facilitating the development of neurodegenerative disorders in humans.

1.2.6. Chronic toxicity observations enable scientists to assess adverse risks of long-term exposures to chemical or physical stressors. LD_{50} data provide a first approximation of potential harm, particularly in the development of pharmaceuticals and pesticides. However, other methodologies are necessary for risk assessment of the vast majority of pollutants in the environment. These protocols generally rely on statistical analysis of epidemiological data.

For example, Bergmans *et al.* (2021) used spatiotemporal and generalized linear models to determine the long-term air pollution effects on depression in older adults. All models included demographic characteristics, individual and community socio-economic status as well as other relevant criteria to assess risks in a cohort of adults from the US Health and Retirement Study.

Mortality may be an end point, for example in cases of prolonged or repeated exposures in subjects compromised by underlying illnesses or lifestyle choices.

Factors relating to bias, confounders, unravelling of exacerbation effects and cause-and-effect issues form the major theme in subsequent chapters.

- Can you suggest why smokers are likely to be more adversely affected by air pollution than non-smokers?

1.3 Exposure Pathways

1.3.1. Select (a), (b) and (d).

Food and water consumption as well as inhalation are the primary routes of entry of xenobiotics. Of particular concern, however, is *in utero* exposure of the fetus to these contaminants, thereby adversely affecting health outcomes in offspring. Similarly, neonates reliant on mother's milk may also receive potentially toxic compounds. During both pregnancy and lactation, pollutants deposited in the mother's depot reserves may re-enter her circulation and, consequently, provide additional toxic burdens for the fetus or neonate.

For a variety of gaseous pollutants and particulates, inhalation into the lungs would be the primary pathway.

Absorption through the skin is another important route of entry into the body. In the case of radionuclides, exposure would occur via multiple pathways, depending on proximity to the source of radiation.

- Can you name the different parts of the lung?

1.4 Interactions

1.4.1. Select (a).

As the term implies, additive effects represent the sum of the individual effects of each component in a binary mixture. In particular settings, populations are often exposed to a complex array of pollutants. The difficulties with ascribing toxic effects to a specific component may, therefore, be amplified in cases where complex interactions occur. Thus, toxic air in polluted cities is composed of gaseous as well as particulate pollutants.

1.4.2. Select (b).

In synergistic interactions, outcomes can be considerably greater than the sum of individual effects of each component in a mixture. When the components act synergistically or via a potentiation mechanism, then risk assessment and prediction of likely end points become markedly more controversial. For example, identifying the synergistic effects of gaseous pollutants and particulates in polluted cities remains a formidable task.

1.4.3. Given the complex array of pollutants in the environment, the adverse impact of interactions in human health outcomes is inevitable, simply on the basis of theoretical considerations. Responses to toxic pollutants may be influenced by:

- within-group interactions, for example relating to gaseous and particulates in vehicular combustion emissions;
- between-group interactions, for example persistent organic pollutants and toxic metals;
- underlying chronic diseases:
  - asthma;
  - cystic fibrosis;
  - chronic obstructive pulmonary disease;
  - cardiovascular disease;
- genetic predisposition; and
- obesity.

Although several of these interactions remain unresolved at present, it is nevertheless instructive and appropriate to consider likely implications for a variety of circumstances and pollutants. Concerns are heightened by a 2020 coroner's ruling that the death of an asthmatic child was due to exacerbations caused by toxic air in London.

- Can you explain 'genetic predisposition'?
- What is asthma?

1.5   Cellular and Metabolic Responses

1.5.1.   The fundamental unit in all living organisms is the cell. It contains the full metabolic machinery for microbes, plants and animals to respond to chemical and physical stressors. The animal cell (Fig.1.1), for example, comprises:

- a nucleus;
- mitochondria;
- endoplasmic reticulum;
- enzymes;
- receptors; and
- cytoplasm, providing a medium for biochemical transformations, including:
    - Phase I reactions;
    - Phase II reactions;
    - Phase III reactions;
    - oxidative stress;
    - inflammation.

Specialized cells develop to form the unique structures of all tissues and organs. In plants, distinctive features include a cell wall, chloroplasts and vacuoles.

The entry of potentially harmful compounds almost invariably provokes a metabolic response in living organisms as pathways are activated to diminish or overcome any adverse effects. The metabolic machinery involved in defence or detoxification processes is quite complex. For some pollutants, the response might entail the use of receptors, signalling pathways and gene expression for the synthesis of the appropriate metabolizing enzymes, usually in the liver of animals.

Biotransformation normally proceeds in three steps. In Phase I, the toxin is modified in a process designed to make it more reactive, for example by hydroxylation catalysed by cytochrome P_{450} mixed-function oxidases.

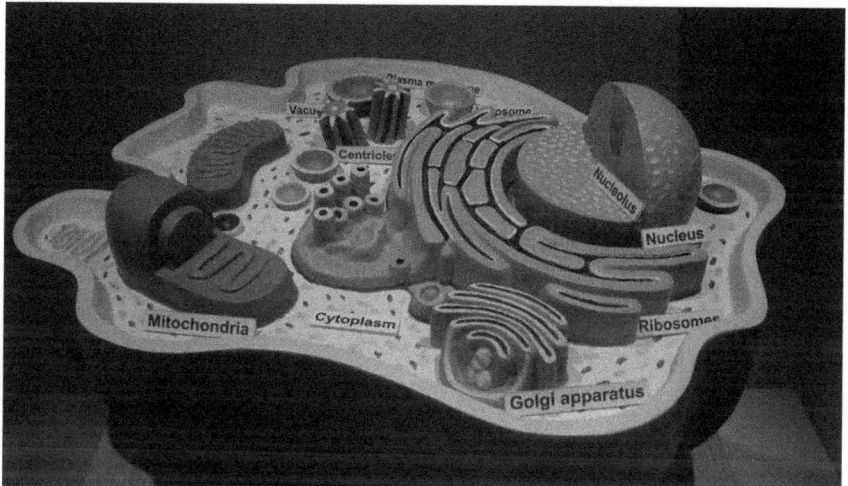

Fig. 1.1. Diagrammatic representation of an animal cell. (File: structure of animal cell. JPG by Royroydeb is licensed under CC BY-SA4.D.)

- Copy this or a similar diagram. Insert and label the 'endoplasmic reticulum'.
- What is the main function of the endoplasmic reticulum?
- In what way is this cell different from:
  ○ a neuron?
  ○ a bacterial cell?
  ○ a plant cell?
- Can you use the article by Lee *et al.* (2021) to summarize the role of mitochondria in animal cells?
- What is:
  ○ a hepatocyte?
  ○ an erythrocyte?

In Phase II, the modified molecule is conjugated with one of a variety of compounds including glutathione, glycine or glucuronic acid.

In Phase III, additional modifications may be introduced and the final product excreted in the urine via the kidneys. However, this process may also lead to the synthesis of reactive metabolites and adducts that may initiate chronic toxicity in the form of cancer and other conditions.

Significant mechanisms exist in higher plants to reduce the impact of different types of environmental stressors. One may involve the mediation of specific amino acids such as proline and histidine. Another mechanism centres on the synthesis of a wide array of peptides, known as phytochelatins, capable of conferring stress tolerance in the plant towards potentially toxic metals.

A mechanism frequently invoked to explain particular features of toxicity in both animals (including humans) and plants centres on the development of 'oxidative stress'. In animals, inflammation is another component of this mechanism. Both oxidative stress and inflammation have been implicated in several age-related and pollutant-exacerbated conditions, including ocular, cardiovascular, pulmonary and neurodegenerative diseases.

- Why are mitochondria so important?
- What is the endoplasmic reticulum?

1.6 Environmental Fate

1.6.1. Knowledge of the environmental fate of pollutants is of critical importance due to potential impacts on human health and biodiversity. Critical factors include:

- intrinsic properties:
  - hydrophilic compounds;
  - lipophilic compounds;
- degradation;
- persistence; and
- bioaccumulation.

Following dispersal into the environment, pollutants may undergo spontaneous or microbial degradation. Alternatively, persistent chemical compounds and plastics may partition into different phases of air, soil, sediments and aquatic systems, with long-lasting implications for humans and wildlife.

The major determinants of pathways and ultimate destination will depend upon the physical characteristics and reactivity of these contaminants in different ecosystems. For example, the environmental fate of hydrophilic and lipophilic compounds will clearly be different.

In aquatic environments, sequential transfer of persistent organic pollutants and metals from prey animals results in excessive accumulations in predators, placing these animals at risk of direct toxicity and/or endocrine dysfunction. Bioaccumulation can also impact on human health, for example in communities reliant on marine products such as blubber which may be contaminated with pesticide residues and toxic metals.

- What is meant by 'hydrophilic'?
- Can you define 'lipophilic'?
- What is 'blubber'?

1.7   Risk Assessment and Regulation

1.7.1.   Following reports of a contamination incident, a series of protocols are generally adopted to assess risks to humans, fragile eco-systems and wildlife. A first step usually involves screening of samples of soil, air, water and food in order to identify the nature, source and extent of pollution.

However, additional biomonitoring steps are necessary to confirm expo-sures to the toxic agent(s) under investigation. Typical bioindicators include levels of the contaminant, enzymes, metabolites and/or hormones in:

- blood;
- milk;
- urine;
- skin;
- stools;
- biopsies;
- organ tissue;
- hair, feathers and integument;
- depot fat;
- eggs;
- blubber; and
- earwax.

The identification of subcellular markers such as enzymes, DNA adducts and signature molecules is also possible with advancements in appropriate methodologies. The enzymes of particular significance in risk assessment are those associated with antioxidant protective mechanisms, including cytochrome P_{450}, superoxide dismutase and reduced glutathione.

Whole organisms can also serve as bioindicators of pollution. The presence of *Escherichia coli* in surface waters is frequently reported by monitoring agencies. The association with sewage contamination means that this bacterium is an indicator of both water quality and food safety. The use of eels has been advocated due to their relatively long lifespan, sedentary existence at certain stages, high fat content and feeding on benthic species.

- What are 'biopsies'?

1.7.2.   A comprehensive network of regulatory agencies has emerged over several decades to monitor and control exposures of humans and wildlife to chemical and physical stressors. This system straddles:

- international organizations:
  - the World Health Organization (WHO);
  - the Food and Agriculture Organization of the United Nations (FAO);
- national agencies:
  - US Environmental Protection Agency (EPA);
- regional authorities:
  - the European Environment Agency (EEA); and
- local jurisdictions.

At the international level, the United Nations environmental provisions are enshrined within the activities of the WHO and the FAO. The renowned national authority is the US EPA which is accepted as a reference benchmark across the globe. In the European Union (EU), the equivalent organization is the European Environment Agency (EEA).

Statutory control and directives apply to virtually all contaminants and pollutants. Notable exceptions include plastics and certain pharmaceutical and personal care products.

Legislation measures generally range from advisories, as for certain mycotoxins, to outright prohibition, as in the case of several persistent organic pollutants. However, there are innumerable examples of contraventions of legal directives. For example, the export of pesticides prohibited by EU legislation has recently been sanctioned by the UK government.

Of immense concern is the WHO statement in 2018 that 93% of all children under the age of 15 inhaled polluted air. This warning remains to be addressed in cities worldwide where WHO guidelines for air quality are routinely ignored. A 2020 survey, for example, indicated that over 75% of polluted UK streets are in London where emissions of toxic pollutants constantly exceed WHO limits.

It is also important to note that a number of heterogeneous groups of compounds identified as 'emerging pollutants' are without regulatory status and not included in surveillance protocols.

- What are pharmaceuticals?
- Why is the term 'emerging pollutants' misleading?

1.8   References

Arumugam, T., Ghazi, T. and Chuturgoon, A.A. (2021) Fumonisin B$_1$ alters global m6A RNA methylation and epigenetically regulates keap1-Nrf2 signalling in human hepatoma (Hep G2) cells. *Archives of Toxicology* 95, 1367–1378. https://doi.org/10.1007/s00204-021-02986-5

Bergmans, R.S., D'Souza, J., Fossa, A., Szpiro, A., Young, M.T. *et al.* (2021) Long-term air pollution exposures and major depression in older US adults: the Health and Retirement Study. *Environmental Health Perspectives. ISEE (International Society for Environmental Epidemiology) Conference Abstracts* 2021(1). Available at: https://ehp.niehs.nih.gov/action/doSearch?Contrib AuthorRaw=Bergmans%2C+Rachel+S (accessed 9 March 2022).

Boente-Juncal, A., Vale, C., Camina, M., Vieytes, M.R. and Botana, L.M. (2020) Re-evaluation of the acute toxicity of palytoxin in mice: determination of lethal dose 50 (LD_{50}) and no-observed-adverse-effect level (NOAEL). *Toxicon* 177, 16–24.

Kergaraval, S.V., Hernandez, S.R. and Gagneten, A.M. (2021) Second-, third- and fourth-generation quinolones: ecotoxicity effects on *Daphnia* and *Ceriodaphnia* species. *Chemosphere* 262: 127823. https://doi.org/10.1016/j.chemosphere.2020.127823

Lee, S., Ko, E., Lee, H. and Shin, S. (2021) Mixed exposure of persistent organic pollutants alters oxidative stress markers and mitochondrial function in the tail of zebrafish depending on sex. *International Journal of Environmental Research and Public Health* 18, 9539. https://doi.org/10.3390/ijerph18189539

2 Biogenic Contaminants

Questions

2.1 Overview

2.1.1. Biogenic contaminants of relevance in environmental toxicology include highly important products of _____ and _____ metabolism.

2.1.2. The three classes of biogenic contaminants that adversely affect environmental and food safety are: _____ _____, _____ and _____.

2.2 Algal Toxins

2.2.1. The incidence of algal blooms in lakes is affected by:

    (a) a preponderance of viruses

    (b) proximity to electronic waste recycling plants

    (c) fertilizers in run-off from arable farms

    (d) climate change

2.2.2. Write a short note on the adverse effects of algal toxins.

2.3 Mycotoxins

2.3.1. Secondary metabolism in fungi is:

    (a) a feature of futile biochemical reactions

    (b) associated with toxin production

© J.P.F. D'Mello 2022. *Key Questions in Environmental Toxicology* (J.P.F. D'Mello)
DOI: 10.1079/9781789248548.0002

(c) a benign activity

(d) of minor ecological significance

2.3.2. Write a short essay on the distribution of mycotoxins that are commonly associated with human health disorders.

2.3.3. Compare and contrast the adverse human health effects of the major mycotoxins occurring in contaminated foods.

2.3.4. Comment on the procedures available to assess and mitigate human health risks associated with exposure to food mycotoxins.

2.4 Phytotoxins

2.4.1. The natural ecology of plants has evolved in response to _____ and _____ factors.

2.4.2. Phytotoxins are:

(a) limited in distribution

(b) of secondary importance in plants

(c) primarily waste products of plant metabolism

(d) critically important as signalling and defence molecules in plants

2.4.3. Elaborate on the significance of the binary activation of named phytotoxins.

2.4.4. Discuss the relative human health risks associated with different food phytotoxins.

Answers

2.1 Overview

2.1.1. Insert: microbial; plant.

The biogenic products of microbes and plants exert significant effects on environmental and food safety, by both physical and chemical mechanisms. Salient issues include:

- structural association between microbial-derived biofilms and plastic polymers;

- pathogen transfer, particle buoyancy and biodegradation of microplastics and adhering co-contaminants;
- adaptation of microbes and plants;
- evaluating the impact of climate change; and
- toxicity in humans and wildlife.

2.1.2. Insert: algal toxins; mycotoxins; phytotoxins.

It is generally assumed that biogenic compounds arise as a result of secondary metabolism in microorganisms and plants. Such an assumption implies subsidiary functions in comparison to the primary metabolism involved in respiration, photosynthesis and protein accretion. However, there is accumulating evidence that biogenic toxins serve crucial roles in defence and competition for survival and growth in particular niches and ecosystems.

A relatively significant number of biogenic compounds are capable of inducing moderate to severe toxicity in humans and wildlife following contamination of foods and drinking water.

- What is 'photosynthesis'?

2.2 Algal Toxins

2.2.1. Select (c) and (d).

The distribution of algal blooms caused by cyanobacteria is increasing at an unprecedented rate in response to climate change, sewage pollution and run-off of farm fertilizers. A recent survey indicated a marked increase in 33 countries. Significant pollution incidents have been reported in:

- Brittany;
- Bolivia; and
- Canada.

Algae emitting hydrogen sulfide contaminating the coastline around parts of Brittany in 2019 contributed to the deaths of three people inhaling the gas. In Bolivia, the first recorded algal bloom occurred in 2015 in a high-altitude lake.

Again in 2015, the Western Lake Erie Basin, situated between the USA and Canada, was subjected to significant nutrient loading from adjacent farmland leading to the formation of the largest algal bloom on record.

- What are the major chemical components of fertilizers used in arable farms?

2.2.2. Algal toxins are associated with a diverse range of adverse effects through contamination in the food chain as well as via direct exposures.

Three classes of toxins have caused particular concern due to deleterious effects on the ecosystem and contamination of marine species, particularly:

- brevetoxins:
    - neurotoxic shellfish poisoning;
- domoic acid:
    - amnesic shellfish poisoning; and
- saxitoxins:
    - paralytic shellfish poisoning.

Algal proliferation in the marine ecosystem frequently results in contamination of seafood with residues of these neurotoxins.

In humans, the brevetoxins are associated with a specific condition known as neurotoxic shellfish poisoning. However, results of a case study suggested that brevetoxin metabolites may be the true cause of neurotoxic shellfish poisoning. Domoic acid is responsible for amnesic shellfish poisoning in humans, while saxitoxins have been linked to the incidence of paralytic shellfish poisoning. The latter condition is characterized by ataxia and respiratory paralysis.

Compounds produced by toxic cyanobacteria include:

- hepatotoxins;
- cytotoxins;
- neurotoxins;
- lipopolysaccharide endotoxins;
- dermatoxins; and
- neuroactive amino acids.

Elevated environmental temperatures especially during the summer season can increase the frequency and number of outdoor recreation activities in rivers, lakes and beaches. This period may also coincide with the occurrence of cyanobacterial blooms. Cases of human dermatitis, eye irritation, diarrhoea and vomiting with allergic-like symptoms following contact with contaminated water have been linked to these blooms.

In 2016, four Florida (USA) counties declared a state of emergency due to flu-like symptoms, respiratory issues, rashes, burning eyes and headaches in people exposed to the aerosols from algal blooms. Inhalation of cyanotoxins is emerging as a significant route of exposure and toxicity.

The likely impact of algal toxins on wildlife should also be considered. For example, in 2020 an alert was issued over predictions of an 'ecological disaster' in Kenyan Rift Valley lakes, following the deaths of over 300 elephants at waterholes, attributed to the incidence of algal blooms. Alkaline and/or stagnant waters as well as pollutants in these and other lakes provide optimum conditions for growth of toxic algae.

Furthermore, marine animals worldwide are frequently at risk to algal toxins in coastal ecosystems. For example, widespread brevetoxin accumulation has been reported in marine mammals along the coast from Texas to Florida (Fire *et al.*, 2021). In contrast, domoic acid exposure has been implicated in severe marine mammal mass mortality events around the US Pacific coast.

2.3   Mycotoxins

2.3.1. Select (b).

Secondary metabolism in many species of fungi is not a benign activity and may exert important ecological functions for certain species of these microbes. For example, some metabolites may serve as infective agents in the induction and/or promotion of fungal diseases of plants.

Those metabolites that are associated with pathological conditions in humans and other animals are termed 'mycotoxins'. Contamination of cereal grains, nuts, fruit and green coffee beans with mycotoxins represents a global food safety issue. When farm livestock are offered feeds containing mycotoxins, then associated residues and metabolites may appear in animal products.

The mycotoxins of particular relevance in human health arise during the secondary metabolism of the following species of fungi:

- *Claviceps*;
- *Aspergillus*;
- *Penicillium*;
- *Fusarium*; and
- *Alternaria*.

Claviceps, *Fusarium* and *Alternaria* are classical examples of toxigenic plant pathogens, while *Aspergillus* and *Penicillium* represent food-spoilage fungi, reflecting postharvest ecology. *Aspergillus fumigatus* is a ubiquitous species, occurring in the soil where it survives and proliferates on organic debris.

It is important to note that toxigenic species of fungi may also directly infect human lungs. Thus, pulmonary infection with *A. fumigatus* is known to exacerbate pre-existing pathological conditions in cystic fibrosis. The condition 'farmer's lung' is another manifestation of direct infection with fungal spores.

- What is 'farmer's lung' disease?

2.3.2. The mycotoxins most commonly associated with human health disorders include:

- ergot alkaloids;
- aflatoxins;
- ochratoxin;
- patulin;
- trichothecenes;
- zearalenone; and
- fumonisins.

The major ergot alkaloids synthesized by *Claviceps purpurea* include the lysergic acid derivatives ergocristine and ergotamine. Aflatoxins B_1, B_2, G_1 and G_2 (AFB_1, AFB_2, AFG_1 and AFG_2, respectively) are synthesized by *Aspergillus flavus* and *Aspergillus parasiticus*. In addition, aflatoxin M_1 may appear in the milk of dairy cows and women consuming and metabolizing AFB_1 from contaminated diets. The two *Aspergillus* species grow and synthesize aflatoxins when temperature and humidity/water activity conditions are favourable.

An outstanding feature in recent surveillance has been the high level of AFB_1 contamination of Indonesian maize, at 428 µg kg^{-1}. Of much concern also, is the relatively high concentrations of total aflatoxins (up to 20 µg kg^{-1}) in maize-based gruels used as weaning food for children in Nigeria. Following an outbreak of human aflatoxicosis in Kenya, 55% of maize products were found with aflatoxin levels exceeding the local regulatory directive of 20 µg kg^{-1}, and 35% containing levels above 100 µg kg^{-1}.

In contrast, *A. fumigatus* occurs widely as an airborne fungal pathogen rather than as a food contaminant. This species produces three major mycotoxins, namely fumitoxin, fumigatin and gliotoxin.

Aspergillus ochraceus and two important *Penicillium* species produce ochratoxin A and ochratoxin B, with the former being more common, occurring with citrinin in cereal grains, dried vine fruits and green coffee. Relatively high values of ochratoxin A in Bulgarian cereals (up to 140 µg kg^{-1}) were associated with grain samples taken from villages with the incidence of Balkan endemic nephropathy. Ochratoxin residues may also occur in products derived from animals fed contaminated grain, with Tangni *et al.* (2021) advocating continued monitoring of meat and edible offal, based on data in Belgium.

A number of *Penicillium* species synthesize patulin and *Penicillium expansum* is of particular significance due to its association with storage rot of apples and other fruits. The presence of patulin in apple juice, attributed to the use of mouldy fruit during processing, has been a cause of concern warranting extensive surveillance in the UK and elsewhere.

The principal *Fusarium* mycotoxins of regular concern include the trichothecenes, zearalenone and fumonisins. Ecological diversity in this group is exemplified by the general association of trichothecenes with cereal grains in temperate latitudes, whereas the fumonisins are common contaminants of maize kernels originating from tropical regions. The trichothecenes include T-2 toxin, HT-2 toxin, diacetoxyscirpenol, deoxynivalenol and nivalenol. Widespread contamination of cereal grains with the trichothecenes and zearalenone has been reported, with some samples exceeding the US advisory limits for deoxynivalenol.

- What is 'nephropathy'?

2.3.3. The diverse chemistry of food mycotoxins translates into complex and often distinct manifestations of adverse effects in humans, as indicated by evidence from prominent case studies. However, with continuing research, common underlying mechanisms are now apparent that provide new and unifying insights into the toxicology of these bioactive compounds. Thus, induction of oxidative stress has been proposed recently as a predisposing event in the toxicity of ochratoxin A and fumonisin B$_1$ (Li *et al.*, 2021). It is likely that the aflatoxins may also cause harm by this mechanism.

In general, mycotoxins induce adverse effects by damaging one or more organs and systems in the body, including:

- liver;
- gall bladder;
- kidney;
- oesophagus;
- immune system; and
- reproductive functions, causing embryo toxicity and other defects.

The liver is the primary target organ for the action of the aflatoxins in humans. The resulting syndrome is referred to as 'aflatoxicosis'. The enduring example of adverse effects is represented by episodes of aflatoxicosis in the tropics. For example, in 1974, an outbreak of liver disease in India was attributed to the consumption of mouldy grain contaminated with aflatoxins. Principal pathological lesions in hepatic tissues included destruction of centrilobular zones, thickening of central veins and cirrhosis. An outbreak of acute hepatitis in Kenya in 1981 was also linked with aflatoxin poisoning, associated with contaminated sources of commercial maize.

The major public health issue, as with other mycotoxins, is the association with carcinogenesis in populations that rely on staples such as peanuts and maize for their nutritional needs. In particular, there is good epidemiological and molecular evidence linking aflatoxin occurrence in food with the incidence of liver cancer. There is also concern that endemic factors such as chronic malnutrition and disease may contribute to, or exacerbate, the development of hepatocellular cancer.

In addition, aflatoxin exposure may enhance the carcinogenic potential of hepatitis B virus. In toxicological classification (International Agency for Research in Cancer (IARC)), AFB_1 has been designated as a Group 1 carcinogen (i.e. sufficient evidence in humans for carcinogenicity), whereas its product in milk (AFM_1) is placed in the Group 2B category (i.e. probable human carcinogen). Furthermore, epidemiological evidence links aflatoxin exposure with other forms of cancer, for example gall bladder malignancy as observed in Chile, Bolivia and Peru.

In contrast, dietary exposure to ochratoxin A has consistently been linked to Balkan endemic nephropathy. This is a chronic disorder occurring among rural populations of Bulgaria, Romania and the former state of Yugoslavia. In affected individuals, the kidneys are markedly reduced in size and, histological examination indicates tubular degeneration and dysfunction, interstitial fibrosis and glomerular defects. However, the co-occurrence of ochratoxin A with citrinin in cereals consumed locally implies an interaction between the two mycotoxins in the aetiology of the disease. A possible endemic ochratoxin-related nephropathy has also been reported in Tunisia. It should be noted that ochratoxin A is classified by IARC as a Group 2B carcinogen.

Other epidemiological evidence indicates a link between dietary fumonisin exposure and human oesophageal cancer incidence in South Africa, China and Iran. Fumonisin B_1 is classified as a Group 2B carcinogen and it has been suggested that it stimulates proliferation of oesophageal cells by modulating the cell cycle and apoptosis. In China, it is suggested that fumonisins may promote liver cancer initiated by AFB_1 and/or hepatitis B virus. Emerging observations point to the carcinogenic potential of ochratoxin A in the liver as well as in the kidney and there may again be interactions with AFB_1.

A consistent theme concerns the identification and characterization of endocrine disruptors. Whereas zearalenone has long been linked with reproductive disorders in experimental models, emphasis is now turning towards the potential of AFB_1 as an antagonist of the androgen biosynthetic pathway. Furthermore, the effects of mycotoxins in immunocompromised patients and resistance to bacterial and viral diseases require elucidation.

Thus, the foregoing account indicates consistent effects of different mycotoxins on carcinogenesis. However, the organs affected are, in general, specific for each mycotoxin. Human health may be further compromised via mycotoxin interactions with immune functions, disease pathogens and endocrine function.

- Can you name a disease that might exacerbate effects of mycotoxins in tropical countries?

2.3.4. There are well-established protocols to assess risks associated with dietary exposures to mycotoxins, based on monitoring of:
- foods:
  - raw ingredients;
  - processed;
- urine;
- stools;
- blood;
- breast milk; and
- organ tissue.

Although analysis of staple foods provides a valuable insight into potential risks from mycotoxin contamination, concentrations in urine, stools, blood, breast milk and organ tissue give a more comprehensive indication of exposures.

Emerging evidence give cause for concern particularly as there may be interactions with other disorders. For example, analysis of urine and stools in Kenya indicated that following feeding of an aflatoxin-free diet, children with kwashiorkor continued to excrete aflatoxins in urine for 2 days, whereas those with marasmus excreted aflatoxins for up to 4 days. Differences were also seen in the type of aflatoxin voided in faeces. Of additional concern is the widespread *in utero* exposure of the fetus to aflatoxins following analysis of cord and maternal blood samples.

Analysis of physiological fluids indicates that human exposure to ochratoxin A is also widespread, with geographical and regional differences in risk. In Croatia, for example, highest blood ochratoxin A levels were observed in subjects living in areas known for the incidence of endemic nephropathy. It is clear now that despite enhanced awareness and the promulgation of advisory and statutory directives, human exposure to mycotoxins continues and not just in developing countries.

Mitigation of risk associated with mycotoxins is only effective through measures to control food contamination, as corrective methods are of limited efficacy. In theory, effective use of fungicides against fungal pathogens of cereal and legume crops should result in reduced mycotoxin contamination of harvested products. However, it is generally accepted that fungicide control is only partially effective and there may be potential risks associated with the development of fungicide resistance in plant pathogens.

Adequate storage of harvested grain, nuts and fruit constitutes a crucial element in the prevention of mycotoxin adulteration, particularly from spoilage fungi. Grain moisture content and environmental temperature are critical factors during storage and transport. In addition, insect and rodent invasion may affect the microclimate within grain silos and also act as significant vectors in transmission of fungal inoculum.

- Can you distinguish between 'cord' and 'maternal' blood?

2.4 Phytotoxins

2.4.1. Insert: intrinsic; extrinsic (environmental).

The evolutionary and ecological success of plants can be attributed to both intrinsic and extrinsic factors. The genetic constitution of plants confers significant versatility in terms, for example, of photosynthesis and nitrogen fixation. These and other constitutive processes provide the basic tools for survival, growth and reproduction.

Important extrinsic factors are climate change, fungal invasion and predation by different species of animals. Plants contain a wide array of physical and chemical deterrents as a defence strategy against such intrusions. The development of thorns in some shrubs and trees is an obvious example of deterrence. However, potential and serious risks to predators occur strategically concealed in the chemistry of plants.

*   Briefly explain the process of 'nitrogen fixation' in plants.

2.4.2. Select (d).

Phytotoxins, also known as allelochemicals, are critically important as signalling and defence molecules in plants. The origin in pathways of secondary metabolism has, hitherto, led to a subjugation of the role of these compounds in plants. However, there is considerable interest in exploiting phytotoxins as biopesticides to replace synthetic, and often hazardous, chemical protectants.

The phytotoxins of particular significance are classified within well-defined groups including:

*   glycosides;
*   non-protein amino acids;
*   phenolic compounds and derivatives;
*   alkaloids; and
*   proteins.

These allelochemicals occur widely in plants of different taxonomic classes. In general, legumes and *Brassica* species contain a more diverse array of phytotoxins compared to cereal cultivars. These compounds occur in all parts of the plant, but the seed is often the most concentrated source, particularly in leguminous species. Tropical plants generally contain a greater diversity of allelochemicals compared to temperate species and the adverse effects in humans may also be striking if the derived food products are not adequately detoxified prior to consumption. In addition, allelochemicals in invasive plants may serve as a competitive mechanism to suppress growth, development and reproduction of other organisms, including forage and agricultural crop species (Krstin *et al.*, 2021).

*   Have you seen any leguminous plants in your garden?

2.4.3. An important feature in plant defence mechanisms is the existence of certain groups of glycosides that induce toxicity via a binary mechanism. The two components in such reactions are stored in separate compartments within the cell and only released upon attack by a predator or fungal pathogen.

It is instructive to consider the role of toxic glycosides due to the diverse range of molecular complexity and mode of deployment of the active principle. Important glycosides include:

26

- cyanogens;
- glycoalkaloids;
- glucosinolates;
- saponins;
- pyrimidine derivatives;
- flavones; and
- ptaquiloside.

The adverse effects of these glycosides are only expressed after completion of an enzyme-dependent reaction releasing the deleterious component from its precursor. This enzyme reaction is triggered by tissue damage to the plant, as, for example, after insect herbivory or fungal penetration.

Cyanogens, for example, exist as a distinctive class of glycosides in the foliage and seeds of different species of plants. The classical example is amygdalin, present in bitter almonds, while another well-known cyanogen, linamarin, occurs in cassava. Following tissue damage and enzyme activation, HCN is released from the cyanogen, causing dose-related toxicity and lethality to insect herbivores and vertebrate animals.

Another example of a binary reaction involves photoactivation of a precursor to release the toxin. The giant hogweed is a dangerous and invasive weed that has altered the ecology of urban areas in Western Europe, the USA and Canada, although it was initially introduced as an ornamental plant. The sap contains compounds known as furanocoumarins that cause phytophotodermatitis in humans resulting in blisters and scars but only when the skin is exposed to sunlight.

The ecological significance of binary activation is that it has the potential to confer on plants a strategic mechanism to deter or resist attack by predators. On the basis of mammalian toxicity, it would be expected that cyanogens may function as highly effective defence compounds, particularly towards arthropod predators. However, the widespread herbivory of cyanogen-containing plants argues against any significant protective attributes.

In contrast, protein phytotoxins may contribute towards a more robust and adaptable system of defence against insect pests (Lin *et al.*, 2021). For example, proteinase inhibitors confer protection to various parts of the plant following wounding by larvae or adult insect pests.

Translation of this knowledge into an over-arching model of plant defence and immunity may contribute to future efforts in reducing the use of harmful chemicals. Realistically, however, the advent of biopesticides remains a distant prospect.

- What is the definition of 'glycoside'?
- What is 'HCN'?

2.4.4. The chemical diversity of food phytotoxins is reflected in the complex range of adverse effects observed in humans. Manifestations include:

- palatability issues and nausea;
- digestive abnormalities; and
- neurological deficits.

The potato glycoalkaloids impart a bitter taste, resulting in palatability issues for consumers. A major outbreak of potato glycoalkaloid poisoning among school children was reported in the UK in 1979. Effects included headaches, vomiting, abdominal pain and diarrhoea. Neurological symptoms ranged from apathy to mental confusion and visual deficits.

Lectins are ubiquitous proteins in the plant kingdom, present in significant quantities in certain seeds and fruits. Kidney bean lectins are widely associated with adverse effects in humans. Particular pathologies arise as a result of inadequate inactivation of the lectins in undercooked beans. Effects include nausea, vomiting, abdominal distension and diarrhoea.

Extensive human health issues continue to be recorded in sub-Saharan communities dependent upon cyanogenic cassava tubers as food. Sub-lethal blood HCN concentrations are regularly observed, leading to symptoms of acute toxicity including manifestations such as irreversible spastic paralysis, known locally as 'Konzo', and cognition deficits. However, age, gender, and protein and micronutrient deficiencies may complicate diagnosis of this condition. Emerging evidence suggests that slower sulfur-mediated cyanide detoxification promotes protein carbamylation. It is suggested that serum carbamylation may be used as marker of motor control and cognition deficits in cassava toxicity (Rwatambuga *et al.*, 2021). An additional complication, highlighted in an alert issued in 2021, is that gut microbes may impact on susceptibility to Konzo poisoning.

The requirement to avoid toxicity of phytotoxins raises questions over constraints regarding processing of foods in developing countries. Particular limitations are associated with inactivation of heat-labile lectins in beans and removal of HCN from cassava, both requiring resources (fuel and water) that are in short supply or unaffordable for communities in sub-Saharan Africa.

• What is cassava?

2.5 References

Fire, S.E., Bogomolni, A., DiGiovanni, R.A., Early, G., Leighfield, T.A. *et al.* (2021) An assessment of temporal, spatial and taxonomic trends in harmful algal toxin exposure in stranded marine mammals from the US New England coast. *PLoS ONE* 16(1): e0243570. https://doi.org/10.1371/journal.pone.0243570

Krstin, L., Katanic, Z., Pfeiffer, T.Z., Marncic, D., Martinovic, A. and Camagajevac, I.S. (2021) Phytotoxic effect of invasive species *Amorpha fruticosa* L. on germination and early growth of forage and agricultural crop plants. *Ecological Research* 36, 97–106.

Li, W., Zhao, H., Zhuang, R., Wang, Y., Rui, R. and Ju, S. (2021) Fumonisin B_1 exposure adversely affects porcine oocyte maturation *in vitro* by inducing mitochondrial dysfunction and oxidative stress. *Theriogenology* 164, 1–11.

Lin, P.-A., Paudel, S., Afzal, A., Shedd, N.L. and Felton, G.W. (2021) Changes in tolerance and resistance of a plant to insect herbivores under variable water availability. *Environmental and Experimental Botany* 183: 104334. https://doi.org/10.1016/j.envexpbot.2020.104334

Rwatambuga, F.A., Ali, E.R., Bramble, M.S., Gosschalk, J.E., Kim, M. *et al.* (2021) Motor control and cognition deficits associated with protein carbamoylation in food (cassava) cyanogenic poisoning: neurodegeneration and genomic perspectives. *Food and Chemical Toxicology* 148: 111917. https://doi.org/10.1016/j.fct.2020.111917

Tangni, E.K., Masquelier, J. and Van Hoeck, E. (2021) Determination of ochratoxin A in edible pork offal: intra-laboratory validation study and estimation of the daily intake via kidney consumption in Belgium. *Mycotoxin Research* 37, 79–87.

3 Ambient Gases and Particulates

Questions

3.1 Overview

3.1.1. Ambient gases that are important in environmental toxicology include: _____ and oxides of ____ and _____.

3.1.2. The _____ and _____ are the principal target organs, although the _____ _____ _____ (CNS) is increasingly implicated in the toxicology of ambient gases and particulates.

3.2 Ozone

3.2.1. Ozone is:

(a) an inert and harmless gas

(b) radioactive

(c) absent from the troposphere

(d) a highly reactive oxidant gas

3.2.2. Give a detailed account of the toxicology of ambient ozone.

3.3 Nitrogen Dioxide

3.3.1. Nitrogen dioxide is:

(a) an inert gas

(b) limited in distribution in the environment

© J.P.F. D'Mello 2022. *Key Questions in Environmental Toxicology* (J.P.F. D'Mello)
DOI: 10.1079/9781789248548.0003

(c) an initiator of free-radical reactions

(d) a catalyst for enzyme activity

3.3.2. Discuss the harmful effects of ambient nitrogen dioxide exposure in humans.

3.4   Sulfur Dioxide

3.4.1. Oxides of sulfur, designated as SO_x, are:

(a) important gaseous pollutants

(b) compounds that cannot be quantified

(c) unable to react with other compounds

(d) excreted from the human body via the urine

3.4.2. Discuss the effects of ambient sulfur dioxide on human health.

3.5   Particulate Matter

3.5.1. Ambient particulate matter is:

(a) a complex mixture

(b) of significance in urban pollution

(c) relevant in rural environments

(d) none of the above

3.5.2. Discuss the human health implications of exposure to ambient particulate matter.

Answers

3.1   Overview

3.1.1. Insert: ozone; nitrogen; sulfur.

The ambient gases of significance in environmental toxicology include ozone, nitrogen dioxide and sulfur dioxide. Satellite images published by the European Space Agency show high levels of atmospheric nitrogen dioxide pollution in China prior to the onset of the coronavirus disease (COVID-19) pandemic in 2019/2020.

A number of questions deserve consideration:

- Is it possible that this pollution exacerbated the severity of COVID-19 symptoms in patients predisposed to respiratory disorders?
- Is there any evidence for such a hypothesis?
- Are there any parallels with the European COVID-19 outbreaks, given the high mortality of patients in polluted cities such as London?

These are questions which are being addressed at research establishments, for example the Martin Luther University in Germany. However, interactions with particulates are also of importance in human morbidity and the detrimental effects may be underestimated.

- Have you seen more recent satellite images of air pollutants?

3.1.2. Insert: lungs; heart; central nervous system.

The lungs and heart are the principal target organs, although the CNS is increasingly implicated in the toxicology of atmospheric pollutants. The role of ambient gases in human morbidity has to be considered in relation to a number of interrelated factors including:

- genetic predisposition;
- physiology;
- lifestyle;
- pulmonary stress;
- cardiovascular disorders;
- neurodegenerative syndromes; and
- carcinogenesis.

Over several decades, air pollution has emerged as the single largest environmental health hazard, causing about 7 million deaths per year worldwide.

- Can you name a cardiovascular disorder?
- What are 'neurodegenerative syndromes'?

3.2 Ozone

3.2.1. Select (d).

Tropospheric ozone is a highly reactive oxidant gas formed by the photochemical reactions involving the following ambient pollutants:

- carbon monoxide;
- nitrogen dioxide; and
- volatile organic compounds (VOCs).

These precursors occur in significant concentrations in vehicular emissions. Existing World Health Organization (WHO) guidelines indicate that any 8-hour period should not exceed a mean of 100 µg m^{-3} of ozone. In urban environments, ozone, together with particulate matter, represent the two dominant air pollutants worldwide.

The American Lung Association recently claimed that almost 40% of Americans reside in areas with potentially harmful levels of ozone. Ozone pollution is also an issue in China where 8-hour maximum concentrations can reach 78 µg m^{-3} in certain regions. This scenario is replicated in polluted cities such as London, Mexico City and New Delhi.

- Outline the photochemical reactions associated with ozone production.
- Name one VOC that might occur in vehicular exhaust emissions.

3.2.2. The toxicology of ambient ozone pollution is established on the basis of:

- epidemiological correlations;
- respiratory tract effects;
- lifestyle factors;
- gender differences;
- genetic polymorphisms;
- activation of alveolar macrophages;
- biophysical interactions;
- inflammatory responses;
- genetic vulnerabilities;
- links to underlying disorders;
- neurological derangements;
- insulin resistance; and
- mortality.

Separately or in combination with other toxic air contaminants, ozone exposure interacts with respiratory and cardiovascular systems, thereby provoking, accelerating onset or exacerbating clinical manifestations of pre-existing disorders.

Extensive epidemiological evidence has demonstrated age, gender and inter-individual differences in the susceptibility to environmental exposures, with specific genetic polymorphisms associated with negative health effects. Thus, the adverse effects of ozone are markedly higher in women than in men, associated with relatively higher fat mass in women. This comparatively larger distribution volume makes women more susceptible to lipophilic contaminants in the environment.

Furthermore, these compounds are also metabolized more rapidly in women, resulting in higher toxicity.

Biophysical interactions are critically important in determining the course of health outcomes. Inhaled ozone damages the nasal and upper respiratory tracts as it travels to the alveoli, the primary sites of adverse effects. Ozone permeating into the alveoli will initially encounter the fluid lining these tissues, known as lung surfactant. A number of constituents of lung surfactant can react with ozone and although some oxidation products will leave the interfacial film, other types such as damaged lipids and surfactant proteins with oxidized amino acid residues may remain in the lung surfactant, thus impairing its ability to function.

In addition, ozone activates the alveolar macrophages and epithelial cells to produce inflammatory cytokines. Short-term exposures cause reduced pulmonary function, airway inflammation and increased bronchial reactivity, which are often reversible. Longer exposures generally result in permanent changes to lung architecture via airway re-modelling and instigation of oxidative stress, resulting in systemic inflammation and extra-pulmonary lesions. Furthermore, high ozone exposures are associated with increased hospital admissions for lung disease.

Children and older adults are particularly vulnerable to the adverse effects of ambient ozone exposure, with pre-existing conditions, reduced immunocompetence and genetic factors adding to the risks. For example, asthmatic patients show greater bronchoconstriction on brief exposure to ozone at 400 ppb.

In chronic obstructive pulmonary disease (COPD) patients, ozone exposure is associated with cardiovascular rather than respiratory events via mechanisms involving increased myocardial energetics and impairment of pulmonary gas exchange. In addition, long-term inhalation of ozone has been associated with systemic oxidative stress leading to endothelial cell activation and pro-coagulation effects, stimulating thrombosis and triggering myocardial infarction, stroke and cardiac arrhythmias. Thrombotic episodes may also occur in healthy individuals exposed to ozone concentrations as low as 0.03 ppm, as demonstrated by vascular markers of thrombosis in blood as well as increased levels of inflammatory indicators such as interleukins and tumour necrosis factor.

The negative impacts of ambient ozone have also been extended to neurological disorders including stroke, Alzheimer's disease, Parkinson's disease and neurodevelopmental abnormalities.

Metabolic effects are an inevitable consequence of exposures to ozone inhalation, as implied in the foregoing account. Specifically, effects on glucose and lipid metabolism may be amplified in diabetics, accompanied by increased insulin resistance (Snow *et al.*, 2021).

In combination with other toxic ambient pollutants, ozone inhalation is consistently linked to premature mortality, principally due to exacerbations of underlying diseases. A 2021 survey indicated that in China, for example, over 93,000 deaths could be attributed to ozone exposure, due to respiratory and cardiovascular impacts.

* Why is insulin important?

3.3   Nitrogen Dioxide

3.3.1.   Select (c).

Nitrogen dioxide is a member of a group of gases known as nitrogen oxides collectively denoted as NO_x. The primary source of NO_x is combustion of fuel in motor vehicles, power generators and off-road installations. Although all NO_x gases are toxic, nitrogen dioxide is considered to be of particular concern for human health due to its properties to:

* initiate free-radical reactions;
* react with unsaturated fatty acids in specific organs;
* induce auto-oxidation of organic components of tissues;
* act at the level of gaseous exchange in the alveoli of lungs;
* damage the surfactant and membrane function in the lungs; and
* increase retention of inhaled particles and potentially interact with other ambient air pollutants.

* What are 'free radicals'?

3.3.2.   The harmful effects of ambient nitrogen dioxide in humans are associated with:

* inflammation and impaired lung functions;
* increased susceptibility to pulmonary infections;
* elevated risks for asthmatics;
* exacerbation of COPD;
* cardiovascular disease;
* interactions with other ambient air pollutants;
* particular risk factors for vulnerable age groups; and
* increased mortality.

Acute exposure to nitrogen dioxide induces lung inflammation characterized by infiltration of serum inflammatory cells and hyperplasia of type-2 respiratory cells. Transcription and release of pro-inflammatory cytokines and tumour necrosis factor occurs, with consequent activation of pulmonary macrophages and lymphocyte proliferation. However, most of these changes have been observed with nitrogen dioxide concentrations exceeding those present in ambient air. In healthy human subjects, short-term exposure to nitrogen dioxide causes elevated levels of natural killer cells, neutrophils and pro-inflammatory mediators. Overall, the responses are more consistent in healthy volunteers exposed to concentrations up to 2 ppm compared to levels below 1 ppm.

Nitrogen dioxide also increases susceptibility to bacterial and viral pulmonary infections. In asthmatics, nitrogen dioxide is associated with increased airway responsiveness to a variety of provocative mediators, including cholinergic and histaminergic compounds, sulfur dioxide and cold air. Exposure to nitrogen dioxide concentrations greater than 0.5 ppm increases the asthmatic response to a range of allergens. According to US Environmental Protection Agency (EPA) health assessments, inhalation of air with high concentrations of nitrogen dioxide over short periods can aggravate respiratory responses in asthmatics, resulting in coughing, wheezing, breathing problems and hospital and emergency admissions.

Moderate increases in airway resistance may also occur in COPD patients following brief exposure to nitrogen dioxide concentrations as low as 1.5 ppm, and decrements in spirometric measures of lung function may become evident with longer exposures to concentrations as low as 0.3 ppm.

Prolonged exposures are associated with cardiovascular effects, incidence of diabetes, deleterious birth end points, cancer and premature mortality. However, there is considerable uncertainty as to whether these effects of nitrogen dioxide are independent of the impacts of other ambient air pollutants such as ozone, sulfur dioxide and particulate matter.

- Can you explain the role of 'macrophages'?

3.4 Sulfur Dioxide

3.4.1. Select (a).

SO_x is the collective expression for a group of important gaseous air pollutants, with sulfur dioxide representing the most important member. Other sulfur gases are less common in the atmosphere. Sulfur dioxide in polluted air is used as an indicator for the entire SO_x category.

Sources of sulfur dioxide are associated with:

- combustion of fossil fuels;
- industrial facilities that utilize it in the manufacture of chemicals such as sulfuric acid, pesticides, preservatives, wine and other beverages; and
- particular countries.

The principal contributors to global sulfur dioxide contamination are China, India, Eastern Europe and the USA.

- Identify the other members of the SO_x group.
- Can you write the formula of sulfuric acid?

3.4.2. Sulfur dioxide inhalation in humans is widely linked to:

- mitochondrial dysfunction;
- airway reactivity and allergenicity;
- enhanced risks for individuals with asthma and COPD;
- occupational exposures;
- deleterious interactions;
- incidence of type 2 diabetes;
- cardiovascular impacts; and
- carcinogenesis.

Effects are discernible at different levels of organization from organelles to tissues and complex physiological systems in a dose- and time-dependent process. For example, sulfur dioxide inhibits expression of key mitochondrial genes, resulting in cellular dysfunction and, ultimately, lung disease.

Inhaled sulfur dioxide reacts with lung surfactant to form sulfurous and sulfuric acids, which together with associated bisulfite and sulfite derivatives are readily absorbed into the bloodstream causing toxicity. Sulfur dioxide can react with other airborne chemical entities, generating particles of, for example, ammonium sulfate.

Sulfur oxide-based particles are capable of being transported into the distal airways and alveoli of the lungs. Following impact with airways, these particles can interact with lung surfactant to activate inflammatory cascades, causing cellular lesions and oxidative stress.

Overall, sulfur dioxide inhalation causes irritation to the nose, throat and the lungs. Asthmatics and individuals with COPD exhibit greater sensitivity to inhaled sulfur dioxide compared to healthy people. Short-term high-level exposures to sulfur dioxide are associated with pulmonary oedema, while short-term low-level exposures can cause bronchoconstriction in asthmatics.

Incidence of increased mortality in episodes such as the London smog of 1952 have focused attention on the role of sulfur dioxide in causing severe health effects, especially in patients with underlying cardiopulmonary conditions. The industrial threshold limit values may not be tolerated by individuals with existing pulmonary disorders such as asthma and COPD.

There is consistent evidence that levels of ambient sulfur dioxide normally tolerated by healthy individuals can cause severe bronchoconstriction in asthmatics. Furthermore, sulfur dioxide in the presence of other pollutants such a nitrogen dioxide may enhance sensitivity to subsequent allergen challenge in these patients. Synergism between sulfur dioxide and cigarette smoking has been established, indicating that the detrimental effects of sulfur dioxide may be increased by other ambient contaminants, including particulates and nitrogen dioxide.

Chronic exposures to ambient sulfur dioxide and ozone have been associated with increased incidence of type 2 diabetes mellitus among adults aged 30–50 years. Whether this occurrence is due to an interaction between the two pollutants cannot be resolved in epidemiological investigations.

Of equal significance is the association between long-term inhalation of sulfur dioxide and the incidence of coronary heart disease in the light of data from a 12-year retrospective cohort study in China. After adjusting for confounding factors, including age, gender, body mass index and lifestyle, risk of this disorder was estimated to increase by 2.5% for every 10 $\mu g \; m^{-3}$ rise in ambient sulfur dioxide concentration (Ma *et al.*, 2021).

Carcinogenic potential represents a further aspect of sulfur dioxide toxicity, particularly following occupational exposure during production or utilization of this gas.

Despite the established effects of sulfur dioxide on human morbidity, synergistic interactions with other ambient pollutants are inevitable, resulting in exacerbation of adverse outcomes. Unravelling the relative impact of each component in this mixture of pollutants remains a challenging issue.

- What is 'allergenicity'?
- Indicate the chemical difference between sulfurous and sulfuric acids.

3.5 Particulate Matter

3.5.1. Select (a), (b) and (c).

Particulate matter is a heterogeneous mixture of primarily microscopic solids dispersed in the gaseous phase of the troposphere, impacting both urban and rural environments. A system of classification has been adopted on the basis of particle size for these aerosols. Emphasis is generally placed on fine particles with a diameter less than 2.5 μm ($PM_{2.5}$) and coarse particles with a diameter less than 10 μm (PM_{10}). Of these, $PM_{2.5}$ is considered to be of greater significance for respiratory impairments and other disorders.

- What is the meaning of 'heterogeneous'?

3.5.2. Ambient particulate matter can impair human health by:

- affecting outcomes in respiratory and cardiovascular diseases. Effects to note include:
  - deleterious interactions with lung surfactant;
  - inflammation of lungs;
  - airway epithelial barrier dysfunction;
  - dysfunction of myocardial mitochondria;
  - cardiac oxidative stress;
- modulating regulatory pathways via complex interactions;
- operating as vectors for potentially harmful contaminants;
- causing gestational risks for mother and fetus; and
- damaging peripheral organ systems.

Respiratory function depends on lung surfactant, a mixture of phospholipids and proteins, secreted by alveolar type II epithelial cells. In common with ambient ozone and nitrogen dioxide exposures, inhalation of particulates is associated with deleterious interactions with pulmonary surfactant. Indeed, Wang *et al.* (2021) concluded that circulating surfactant proteins may serve as biomarkers of respiratory injury caused by ambient particulate-matter pollution. There is also evidence that particulates may directly contribute to airway epithelial barrier dysfunction by affecting regulators of paracellular permeability.

The foregoing observations are consistent with extensive epidemiological evidence of complex associations between the incidence of respiratory and cardiovascular disease, including deaths, following both short-term and long-term exposures to $PM_{2.5}$ emissions. Patients with underlying illnesses such as asthma and COPD are particularly vulnerable to the adverse effects of particulates. For example, 2021 data indicated that ambient ultrafine particulates were associated with increased paediatric emergency-department attendance for main respiratory diseases in Shanghai (China). The dose–response relationship was almost linear, with asthma incidence as the prominent reason for hospital admissions.

Emerging observations indicate that respiratory health in the elderly is affected to a greater extent by ultrafine and fine particulates, relative to PM_{10} pollution. Other 2021 observations confirm the independent association between increased cardiovascular mortality and short-term exposure to ambient particulates even at concentrations below current advisory and regulatory limits.

The pathways to toxic impacts in humans are depicted in Fig. 3.1. Emerging evidence indicates that particulates are toxic due not only to direct impacts associated with physical properties, but also to roles as vectors of polycyclic aromatic hydrocarbons (PAHs) and metal contaminants. The net effect of these combinations is the generation of reactive oxygen species (ROS) and inflammatory processes, depending on the redox status of individuals exposed to ambient contaminants in urban conurbations.

Inflammation of the lungs may release inflammatory mediators (oxidants and cytokines) which enter the circulation, thereby promoting inflammatory reactions in the walls of blood vessels and endothelium. Another mechanism involves the direct transport of nanoparticles in the blood to the cardiovascular epithelium. Uptake of nanoparticles by cells of the immune system and the autonomic reflex pathways, specifically affect cardiac functions including heart rate variability.

Equally, there is emerging evidence that nanoparticles can induce dysfunction of myocardial mitochondria and cardiac oxidative stress. Adverse biochemical effects are likely to be wide ranging in view of the critical functions of mitochondria (Fig. 1.1) in bioenergetics, redox balance and regulation of cell signalling. It has been suggested that iron-rich airborne ultrafine particles in urban air constitute a plausible risk factor for development of cardiovascular disease at all stages of life from prenatal through to adulthood due to the induction of mitochondrial deficits (Maher *et al.*, 2020).

Of considerable significance is the recent observation that air pollution nanoparticles have been detected on the fetal side of placentas, presenting greater risks for pregnant women living near congested roads.

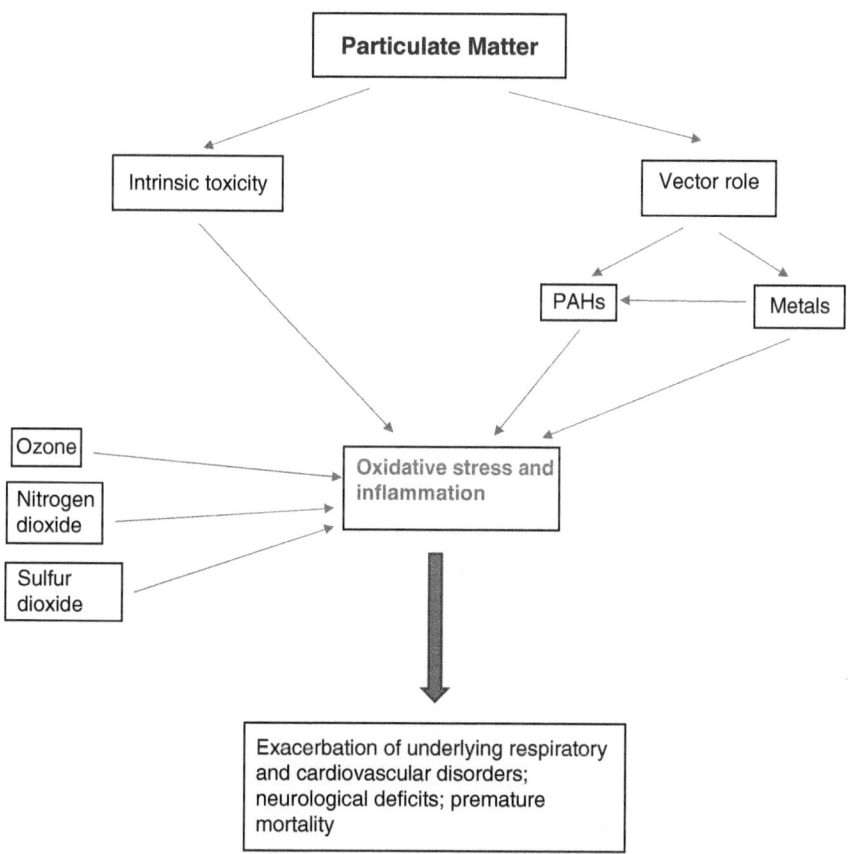

Fig. 3.1. A model illustrating the dynamics and dual roles of ambient particulate matter in human morbidity. The effects of extraneous pollutants, including polycyclic aromatic hydrocarbons (PAHs) and metals are evoking considerable toxicological interest even at the molecular level. The relationship to co-occurring gases in urban pollution is also shown to present an integrated picture of the complex interactions contributing to human health disorders.

Emerging evidence also links exposure to particulates as well as NO_x to incidence of age-related macular degeneration and retinal thickness. Although age and genetics are well-known causal factors for these end points, smoking tobacco adds significantly to the over-all risk. It is plausible, therefore, that ambient air pollution should be considered as an avoidable risk factor for ocular damage, given also that the retina is one of the highest oxygen-utilizing tissues in the body.

- Explain the term 'redox'.

3.6 References

Ma, Z., Zhuang, Z., Cao, X., Zhang, L., Liu, Y. *et al.* (2021) Association between long-term exposure to sulfur dioxide pollution and incidence of coronary heart disease in northern China: a 12-year retrospective cohort study. *Atmospheric Pollution Research* 12(4), 60–65. https://doi.org/10.1016/j.apr.2021.02.006

Maher, S.A., González-Maciel, A., Reynoso-Robles, R., Torres-Jardón, R. and Calderón-Garcidueñas, L. (2020) Iron-rich air pollution nanoparticles: an unrecognized environmental risk factor for myocardial mitochondrial dysfunction and cardiac oxidative stress. *Environmental Research* 188: 109816. https://doi.org/10.1016/j.envres.2020.109816

Snow, S.J., Henriquez, A.R., Fisher, A., Vallanat, B., House, J.S. *et al.* (2021) Peripheral metabolic effects of ozone exposure in healthy and diabetic rats on normal or high-cholesterol diet. *Toxicology and Applied Pharmacology* 415: 115427. https://doi.org/10.1016/j.taap.2021.115427

Wang, Z., Xu, M., Wang, Y., Wang, T., Wu, N. *et al.* (2021) Air particulate matter pollution and circulating surfactant protein: a systemic review and meta-analysis. *Chemosphere* 272: 129564. https://doi.org/10.1016/j.chemosphere.2021.129564

4 Persistent Organic Pollutants

Questions

4.1 Overview

4.1.1. In environmental toxicology, persistent organic pollutants (POPs) are conventionally defined as compounds _____ to degradation in different ecosystems.

4.1.2. An overriding feature of POPs is their _____ properties, favouring accumulation in specific body compartments, such as adipose tissue.

4.1.3. Comment briefly on the diverse properties and environmental impact of POPs.

4.2 Polycyclic Aromatic Hydrocarbons

4.2.1. Polycyclic aromatic hydrocarbons (PAHs) are:

    (a) aliphatic in terms of chemistry

    (b) unrelated to each other

    (c) a group of structurally related compounds

    (d) limited in environmental distribution and impacts

4.2.2. Discuss the human health risks associated with exposure to PAHs, commenting also on possible mechanisms underlying adverse effects.

4.3   Polychlorinated Biphenyls

4.3.1. Polychlorinated biphenyls (PCBs) are:

(a) polysaccharides with chlorine substitutions at key points in the structures

(b) a group of compounds containing phosphorus, cadmium and boron

(c) radioactive

(d) a group of structurally related synthetic compounds

4.3.2. Evaluate the toxicological evidence for current human health concerns over exposures to PCBs in the environment.

4.4   Dioxins and Furans

4.4.1. Dioxins and furans are:

(a) simple aliphatic molecules

(b) polychlorinated compounds

(c) synthesized by toxic algae

(d) examples of PAHs

4.4.2. Discuss the toxicology of dioxins and furans with particular reference to human health risks.

4.5   Insecticides

4.5.1. Organochlorine and organophosphorus insecticides are:

(a) compounds with similar chemical structures

(b) lipophilic

(c) soluble in water

(d) toxic to non-target species

4.5.2. Compare the toxicological properties of the three major classes of organochlorine insecticides, with particular reference to human health risks.

4.5.3. Submit an essay on the toxicology of organophosphate insecticides, emphasizing human health risks.

4.6 Fungicides

4.6.1. Fungicides currently in use:

(a) belong to a uniform class of structurally related compounds

(b) always contain sulfur

(c) are primarily organic compounds

(d) only act on target pathogens

4.6.2. Briefly comment on the toxicological issues associated with fungicide exposures in humans.

4.7 Herbicides

4.7.1. Herbicides currently available:

(a) never contain phosphorus

(b) are non-toxic to humans

(c) act on biochemical processes unique to plants

(d) may contain chlorine

4.7.2. What evidence is there to support the opinion that herbicides are toxic to humans?

4.8 Pesticide Resistance

4.8.1. Discuss the mechanisms underlying the development of insecticide resistance in target species.

4.8.2. What is the potential impact of the development of fungicide resistance in plant pathogenic fungi?

4.8.3. Submit an essay entitled: 'Herbicide Resistance'.

4.9 POPs as Endocrine Disruptors

4.9.1. Endocrine disruption in humans and other vertebrate species is primarily associated with:

(a) phosphorus in the environment

(b) endogenous urinary nitrogen

(c) endocytosis

(d) diverse organic compounds in the environment

4.9.2. Discuss the likely impact of environmental endocrine disruptors on human physiology and health.

4.10 Ecotoxicity

4.10.1. Explain how POPs may affect biodiversity in the ecosphere.

Answers

4.1 Overview

4.1.1. Insert: resistant.

The term 'persistent organic pollutants' (POPs) is a collective name for recalcitrant molecules. POPs are particularly resistant to degradation in biological systems. For this reason, several members of this group are also classified as 'forever chemicals' in the popular media.

POPs considered to present specific and continuing risks for humans and wildlife species include:

- polycyclic aromatic hydrocarbons (PAHs);
- polychlorinated biphenyls (PCBs);
- dioxins;
- organochlorine pesticides;
- organophosphate pesticides;
- fungicides; and
- herbicides.

These chemical compounds are distinct from another group of commonly occurring pollutants known as volatile organic compounds (VOCs). These include the BTEX sub-group comprising benzene, toluene, ethylbenzene and xylene. Acetaldehyde is another member of the VOC group. VOCs are generally associated with vehicle emissions and may act separately or synergistically with other ambient pollutants.

- Draw the structure of benzene.

4.1.2. Insert: lipophilic.

POPs are lipophilic, accumulating in the fatty tissues of vertebrate species, including humans, following exposure via inhalation, food or water. These accumulations may appear in different organs and tissues including the liver, muscle and subcutaneous tissue. Such deposits may serve as biomonitoring indicators of exposure. For example, a study by Quinete *et al.* (2020) demonstrated the presence of organochlorine compounds and PAHs in liver and muscle of juvenile Magellanic penguins stranded off the south-east coast of Brazil.

- Can you define 'subcutaneous'?

4.1.3. As a disparate group of contaminants, POPs are unified in shared properties of diverse biophysical as well as physiological effects. Important issues include:

- chirality;
- structure–activity relationships;
- legacy impacts;
- diverse toxic manifestations;
- complex interactions; and
- biomagnification.

Chirality appears to be a common feature of several POPs which may modulate metabolic impacts on cellular enzymes, transporters and particularly DNA.

Specifically, levels of chiral signatures of persistent organochlorine compounds have been determined in human tissues. Similarly, the enantioselective occurrence of chiral PCB congeners has been measured in cetaceans from the Mediterranean Sea. Meanwhile, other data indicate the comparative aquatic toxicity of current chiral pesticides. Kallenborn *et al.* (2021) recommend that chirality should be an integral component of environmental risk assessments for compounds such as agrochemicals and pharmaceuticals.

The structure–activity properties of POPs and interactions with cellular receptors determine overall outcomes for environmental fate and metabolic impacts. An overriding feature of POPs is their resistance to breakdown by natural processes in different ecosystems and their lipophilic properties which promote accumulation in animal products and human milk, and, consequently, present risks for breast-fed infants and those receiving cow's milk.

Although classified together, POPs are known to elicit widely different toxicities in humans and other animals, potentially threatening the very existence of some species. Considerable concern is now emerging over the effects of POPs as endocrine disruptors.

In addition, biomagnification of certain POPs in the trophic chain is presenting particular risks for endangered ocean predators. Interaction with other environmental stressors such as microplastics in marine ecosystems is another factor of concern for certain POPs.

- Did you learn about 'chirality' in your A-level chemistry course?
- Are you familiar with the 'biomagnification' trophic sequence?

4.2   Polycyclic Aromatic Hydrocarbons

4.2.1. Select (c).

PAHs are a diverse group of structurally related compounds comprising fused aromatic rings, ranging in complexity and including:

- naphthalene, the simplest, containing two fused rings;
- anthracene, a three-ringed compound;
- pyrene, chrysene and naphthacene, composed of four rings in different structural arrangements;
- five-ringed PAHs such as benzo(a)pyrene, benzo(e)pyrene and perylene; and
- numerous six-ringed compounds all with different structural configurations, including linear, cyclical, alternant, non-alternant, branched and non-branched conformations.

- What are 'fused aromatic rings'?

4.2.2. Public health risks associated with PAHs may occur after:

- dermal contact;
- inhalation; and
- consumption of contaminated foods.

Differences in exposures appear to be dependent upon residential location. In the Niger Delta of Nigeria, for example, 2021 data imply that dermal contact might well contribute to public health risks for local residents due to contaminated marine sediments.

Elsewhere, communities in polluted cities will be exposed to petrogenic sources from vehicular and industrial sources. Inhalation is, therefore, a major route for entry into the body. Similarly, homeowners using wood burners or those residing in the vicinity of forest fires will be exposed to pyrogenic sources, again with inhalation as the primary route of entry. In contrast, indigenous communities in the Bigstone Cree Nation in Alberta (Canada) may consume traditional foods contaminated with PAHs that are released during the industrial extraction of oil sands at nearby installations (Golzadeh *et al.*, 2021).

It is not possible to link the foregoing risk assessments to specific cases of human morbidity, despite evidence of well-established manifestations of toxicity in animal models. These include reproductive dysfunction, cardiovascular disorders, bone marrow abnormalities, immune suppression and hepatic lesions.

In addition, a wide range of PAHs are associated with carcinogenesis in animal models. Human health concerns over PAHs have, therefore, centred on pathways to carcinogenesis as determined by:

- biotransformation;
- metabolism;
- adduct formation; and
- structural properties.

It is known that PAHs undergo transformations in living organisms which determine the ultimate biochemical fate towards excretion or malignancy. In the latter pathway, activated metabolites form covalent adducts with DNA, an initial step in chemical carcinogenesis.

As a general rule, PAHs with fewer than four fused rings are non-carcinogenic, whereas those with six rings are mostly carcinogenic. The variation in carcinogenic potential can be explained on the basis that some characteristics are necessary in both the parent molecules and in their respective metabolites to form DNA adducts. In PAH topology, some regions and carbon-atom positions determine biological activity. For example, the presence of the bay region in PAHs is generally considered to be a major prerequisite for carcinogenic activity.

The International Agency for Research in Cancer (IARC) classifies carcinogenicity of pollutants and other potentially deleterious compounds within four groups. Regarding PAHs, the evidence is based primarily on studies with animal models. Most PAHs are designated as Group 3 (i.e. unclassifiable as to carcinogenicity in humans) whereas three, namely benz(*a*)anthracene, benzo(*a*)pyrene and dibenz(*a,h*)anthracene are considered to be probable human carcinogens (Group 2A). A further nine PAHs appear in Group 2B as possible human carcinogens; two of these include dibenzo(*a,e*)pyrene and dibenzo(*a,h*)pyrene.

Thus, it is clear that carcinogenicity is the dominant issue in concerns over the human health effects of PAHs. Genetic and lifestyle factors such as cigarette smoking may exacerbate outcomes following exposures to PAHs.

- What is an 'adduct'?

4.3   Polychlorinated Biphenyls

4.3.1. Select (d).

PCBs are a member of a relatively large class of structurally related organic compounds. Salient characteristics and functional properties to note are:

- existence of 209 possible PCBs, known as a congeners;
- that variants are classified into two categories: dioxin-like and non-dioxin-like, based on structural differentiation;
- that the same physicochemical properties that conferred desirable industrial and commercial applications ensured that PCBs would be highly resistant to degradation in the environment;
- that PCBs are highly lipophilic and thus bioaccumulate up the trophic system, concentrating in fatty tissues; and
- emerging data suggest that non-dioxin-like PCBs predominate in diverse matrices, including human tissues.

- Are you familiar with the structure of 'biphenyls'?

4.3.2. The toxicological basis for continuing human health concerns over exposures to PCBs centres on:

- bioaccumulation;
- toxicity profiles;
- *in vivo* research assays;
- population-based risk assessments; and
- specific case studies.

Following entry into the human body, PCBs accumulate in lipid fractions of adipose, liver and brain tissues, but are also present in quantifiable levels in blood. Experimental data demonstrates that PCBs can cross the placenta and also accumulate in breast milk. An analysis of breast milk samples in Canada during the early 1970s indicated contamination with PCBs in almost all samples.

Until recently, most adverse human health effects have been attributed to dioxin-like PCBs. Apart from structural resemblance to dioxins, this class of PCBs, like dioxins, bind to the aryl hydrocarbon receptor (AhR). Consequently, the toxicity profiles of dioxin-like PCBs are similar to those of dioxins. Thus, exposure to high levels of dioxin or dioxin-like PCBs can precipitate chloracne and liver damage, while chronic exposure to lower levels may compromise immune function or cause cancer.

In contrast, the non-dioxin-like PCB congeners have negligible AhR-binding activity. Consequently, it was long assumed that this class of PCBs were toxicologically benign. However, subsequent work demonstrated that certain non-dioxin-like PCBs, but not dioxin-like PCBs, interfered with dopamine signalling and altered Ca-dependent signalling in neurons *in vitro*. These observations corroborated earlier epidemiological and preclinical studies with animal models indicating the neurotoxic nature of at least some PCBs and that the developing nervous system is considerably more sensitive than that of adults with mature networks. These initial developments were given added significance following data from the US Centers for Disease Control confirming widespread exposure to PCBs among women of childbearing age living in the USA.

Research evidence obtained with *in vivo* models has also contributed ongoing health concerns over human exposures to PCBs. Despite questions over interpretation of data, it is possible to discern reliable toxicological impacts of exposures to PCBs including:

- reproductive abnormalities such as depressed conception rates, increased fetal mortality and reduced birthweights of progeny;
- oestrogenic modulation;
- teratogenic effects;
- reduced plasma thyroxine status, increased circulating thyroid-stimulating hormone and altered features of thyroid histology;
- impaired neurological development causing behavioural abnormalities;
- compromised immune function; and
- carcinogenic potential.

Contemporary data confirm that human exposure to PCBs occurs primarily through intake of contaminated food and indoor inhalation. However, Weitekamp *et al.* (2021) identified important limitations in the data available to satisfactorily assess population exposures to PCBs. Nevertheless, other researchers claimed in 2021 that non-dioxin-like PCBs found in environmental and human serum samples may contribute to neurotoxicity in individuals consuming fish and in children inhaling indoor air in schools.

Evidence from case studies has revealed both non-cancerous and malignant outcomes in humans exposed to PCBs, broadly consistent with the *in vivo* data outlined above. Key manifestations worth noting are:

- chloracne;
- hyperpigmentation;

- ocular deficits;
- reproductive abnormalities;
- neurotoxicity; and
- classification as Group 1 carcinogens (IARC).

Substantial data has accumulated following the accidental exposure of people in Japan and Taiwan consuming rice oil inadvertently contaminated with PCBs. Subsequent to these incidents, affected individuals developed chloracne and hyperpigmentation of the skin, gingiva and nails, with partial recovery over time, although symptoms were still present 10–14 years later. Ocular lesions including hypersecretion and swelling of the sebaceous glands of the eyelids are also common with PCB intoxication.

Other findings indicate that women in a number of high-exposure environments, such as where there have been industrial or accidental discharges, have given birth to offspring with lower birthweights or shorter body lengths and/or smaller head circumferences.

The neurological implications of exposure to PCBs have been assessed in mothers who ingested contaminated rice oil in Japan and Taiwan. It was considered that prenatal exposure to PCBs and other POPs may adversely affect the neurological development of children born to these mothers. The evidence showed that infants of the exposed mothers exhibited a range of neuropsychological deficits which persisted for several years, but the effects of other POPs present as co-contaminants could not be excluded in this analysis.

The Taiwan incident also suggested other toxicological effects in victims, including: (i) substantial elevation in mortality rates due to cirrhosis and chronic liver disease; (ii) an increased incidence of goitre in both men and women; (iii) immunological defects; and (iv) occurrence of bronchitis-like syndromes, marked by a large quantity of expectorant during the initial phase of exposure.

The carcinogenic potential of PCBs has been a cause of considerable concern, as with other POPs. Following the Japanese poisoning episode, a significant increase in the incidence of mortality attributable to cancer of the liver and respiratory system occurred in men but not in women. Based on extensive animal studies and epidemiological observations, the IARC has classified PCBs as Group 1 human carcinogens.

Although the evidence reviewed above relates primarily to PCBs, it should be recognized that interactions may, and often do, occur with other pollutants, for example dioxins. Thus, 2021 data indicate possible associations of PCBs, dioxins and furans with follicle-stimulating and luteinizing hormones in postmenopausal women participating in the US National Health and Nutrition Examination Survey covering the period 1999–2002 (Lambertino et al., 2021). Interactions in carcinogenesis are also likely to occur in other situations. Consequently, the human toxicology of PCBs cannot be regarded as closed.

- Can you recall the functions of 'follicle-stimulating' and 'luteinizing' hormones?

4.4 Dioxins and Furans

4.4.1. Select (b).

Dioxins and furans are polychlorinated organic compounds with particular structural features. The terms dioxins and furans are abbreviations for polychlorinated dibenzo-p-dioxins (PCDDs) and polychlorinated dibenzofurans (PCDFs), respectively. PCDDs and PCDFs represent two groups of planar, tricyclic compounds which can accommodate up to eight Cl atoms attached to specific carbon atoms. In total, there are 75 possible PCDD congeners and 135 possible PCDF analogues.

Noteworthy properties include:

- lipophilicity;
- persistence;
- pathways in pollution;
- exposure routes;
- metabolism; and
- bioaccumulation.

Similar to PCBs, dioxins and furans are lipophilic, resistant to biodegradation and, therefore, persistent in the environment, attaining notoriety in the Vietnam War particularly during the period 1961–1971. Despite the passage of time (c.50 years), dioxin levels remain high in soils and sediments contaminated with Agent Orange by the US Air Force. Dioxins and furans are undesirable by-products of many chemical industrial activities and of all combustion processes.

- Can you draw a 'tricyclic' structure?

4.4.2. The toxicology of dioxins and furans continues to evolve in several respects, in the knowledge of extensive exposures in humans. Legacy and accidental exposures add to ongoing toxicological concerns. Emerging evidence and alerts consistently emphasize:

- diverse effects in animal models paralleling manifestations in human morbidity;
- risks for infants, older children and lactating mothers;
- accidental and occupational exposures associated with long-term effects;
- carcinogenicity;
- metabolic stability;
- potential links to incidence of diabetes;
- complex mechanisms via the aryl hydrocarbon receptor (AhR); and
- interactions between AhR and vitamin A metabolism.

Manifestations of dioxin toxicity in animal models include:

- skin defects;
- liver damage;
- immune suppression;
- carcinogenesis;
- abnormalities in fetal development; and
- endocrine disruption.

These diverse effects and widespread environmental distribution of dioxins have ensured an active and continuing agenda for surveillance of human exposures and toxicity.

The risk for infants has been assessed from determination of dioxins in breast milk. The results of a recent study in China indicated that mothers and their breastfeeding infants residing near an e-waste recycling site were exposed to higher levels of PCDDs/PCDFs than those in a control group (Luo et al., 2021).

In Ghana, 2021 data indicate lower exposure levels for breastfeeding infants than those in industrialized countries. However, the estimated intakes of dioxins and furans still exceeded the safety limits issued by the Agency for Toxic Substances and Disease Registry and the World Health Organization (WHO).

In children, in utero exposure to dioxins and/or PCBs at or near background levels has been linked with effects on neurodevelopment, behaviour and thyroid hormone status. At higher levels, children exposed transplacentally to these contaminants exhibit:

- skin abnormalities such as chloracne;
- developmental deficits;
- low birthweight;
- behavioural disorders;

- reduced height among girls at puberty; and
- hearing deficits.

It is not fully known whether these effects are caused by dioxins and related chemicals or other contaminants in the environment, but there is a general consensus that subtle effects might already be occurring in the general population at current background levels of exposure to these contaminants.

There are a number of cohorts with high exposure to PCDDs/PCDFs (including PCBs and additional contaminants) associated with the Vietnam conflict (particularly the period 1961–1971) and the Seveso chemical plant explosion in Italy in 1976 or in occupational circumstances. In one such population highly exposed for more than a year and with a 20-year latency period, there was an increase in all cancers. The Vietnam investigation showed higher incidence of diabetes correlating with increasing dioxin levels, but no other effects. In the Seveso survey, residents had elevated dioxin levels and there were significantly more girls born than boys, indicating a change in normal sex ratio in that population.

Four epidemiological studies of high-exposure industrial cohorts in Germany, the Netherlands and the USA indicated higher overall cancer mortality rates. In general, the strongest evidence for the carcinogenicity of 2,3,7,8-tetrachlorodibenzo-ρ-dioxin (TCDD) is for all cancers combined, rather than for any specific type. In these cohorts, blood lipid levels of this dioxin were markedly elevated compared to values for populations exposed to background levels.

In the IARC classification, 2,3,7,8-TCDD is designated as carcinogenic to humans. This conclusion was based on a number of factors including the observation that 2,3,7,8-TCDD is a multisite carcinogen in animal models, involving AhR, and that this receptor is highly conserved in an evolutionary and functional context in humans and animal models. Furthermore, tissue concentrations of this dioxin were comparable in humans and a rat model. Other PCDDs and non-chlorinated congeners are not classifiable as to their carcinogenicity in humans.

In addition, the IARC concluded that there was inadequate evidence for the carcinogenicity of PCDFs in humans, although there is limited evidence for carcinogenicity of certain congeners in animal models. The US Environmental Protection Agency (EPA) confirmed the IARC classification for TCDD as a 'human carcinogen' and thus satisfies the stringent criteria required to accept a causal relationship between TCDD exposure and cancer risk.

Despite the foregoing reservations, there can be no doubt as to the carcinogenic potential of dioxins individually or as a group or in combination with PCBs, in view of the incidence of mortality and malignancy following high-exposure accidents in industrial environments. Furthermore, combinations of chlorinated solvents and dioxins/furans, from industrial sources, have been implicated in the incidence of thyroid cancer for the general population.

The 1968 Yusho incident in Japan sheds further light on dioxin-related health hazards. Over 2000 individuals were intoxicated by high concentrations of dioxins and PCBs in contaminated edible rice bran oil. Symptoms displayed in this population included:

- general malaise;
- headache;
- abdominal pain;
- paresthesia;
- swelling of joints; and
- cough.

These disorders were collectively known as 'Yusho'. A 2021 study has demonstrated that blood concentrations of dioxins and furans remain elevated in survivors 50 years after initial onset of toxicity. It is clear that these congeners are biologically stable and resistant to cytochrome P_{450} metabolism

The relationship of dioxins and furans to diabetes and associated nephropathy is emerging as an additional feature of risk in humans. These associations have been inferred from meta-analyses which indicate that repeated exposure to TCDD appears to be important when considering associations with diabetes. Of eight dioxin-like compounds, six were linked with diabetes, five with diabetes without nephropathy and seven with diabetic nephropathy. The underlying mechanisms of these associations have yet to be established. However, results of a 2021 study indicated that beta cells in isolated pancreatic islets are not only sensitive to but also a specific target of the toxic action of TCDD.

The AhR is frequently invoked as the focal point for the action of POPs, including dioxins and furans. Results of a 2021 study indicate a role for AhR in overall retinoid metabolism. It was argued that pathological lesions following AhR over-activation by dioxins were strikingly consistent with manifestations associated with vitamin A deficiency or excess including:

- disturbance in bodyweight regulation;
- fertility deficits;
- spermatogenesis abnormalities;
- fatty liver;
- cirrhosis;
- hepatocellular carcinoma;
- bone lesions; and
- developmental malformations.

It was concluded that AhR exerts a pivotal role in dioxin-induced retinoid disruption.

Thus, there appears to be a justifiable interest in the diverse effects of dioxins and furans in human morbidity. The metabolic mechanisms involving AhR, however, remain a nebulous issue.

- What is 'vitamin A'?

4.5 Insecticides

4.5.1. Select (b) and (d).

Organochlorine and organophosphorus insecticides belong to two distinct groups of lipophilic compounds with diverse chemical structures. Both groups comprise compounds that are highly toxic to non-target species, including humans.

Organochlorine insecticides are a heterogeneous group of compounds consisting of three different chemical classes:

- diphenylethanes – these include:
    - dichlorodiphenyltrichloroethane (DDT);
    - dicofol;
    - methoxychlor;
- cyclodienes – important examples are:
    - aldrin;
    - endrin;
    - dieldrin;
    - chlordane;
    - heptachlor;
    - endosulfan;
    - toxaphene; and
- cyclohexanes – these include a variety of isomers such as:
    - lindane;
    - β-hexachlorocyclohexane.

These and other classes of insecticides were designed primarily to control insect-borne human diseases and agricultural pests.

Organophosphate insecticides comprise a wide range of compounds with diverse chemical structures. Major classes include:

- phosphates – this group includes:
    o dichlorvos;
- thionophosphates – these comprise:
    o chlorpyrifos;
    o fenthion;
    o parathion; and
- dithiophosphates – this group includes:
    o malathion.

- Can you name an important insect-borne disease affecting humans?

4.5.2. Organochlorine insecticides are associated with diverse toxicological effects based on:

- exposure routes;
- modulation of neuronal sodium channels;
- disruption of neurotransmitter signalling;
- metabolic dynamics;
- epidemiological links to neurodegenerative disorders;
- developmental and lactational risks;
- induction of oxidative stress; and
- endocrine perturbation.

Human exposure to organochlorine insecticides occurs via different routes, including dermal, inhalation and oral. There may be additional exposures for the developing fetus and neonates. In general, dermal exposure to dichlorodiphenyltrichloroethane (DDT) is well tolerated and does not result in adverse effects. In contrast, acute oral exposures are most notably accompanied by hypersensitivity in the buccal cavity, followed by spontaneous motor movements, muscle hyperexcitability, tremors and cognitive aberrations.

The said neurological manifestations have been attributed to the well-defined mechanism of action of DDT specifically directed at the function of the sodium channels in neurons. The brain and other regions of the central nervous system are important target organs for the toxicity of organochlorine insecticides. DDT binds to the sodium channels, thereby disrupting dynamics of sodium and potassium ions and resulting in neuronal hyperexcitability.

Furthermore, DDT can disrupt neurotransmitter signalling by affecting glutamate flux in the brain and altering the levels and functioning of serotonin, norepinephrine and dopamine in different regions of the brain. These perturbations are of considerable significance as DDT and its metabolite dichlorodiphenyldichloroethylene (DDE) have been identified in post-mortem brain tissues of individuals with neurodegenerative disorders such as Parkinson's disease (PD) and Alzheimer's disease.

Additional risks are associated with the relative ease with which DDT is transported around the body, particularly from mother to the developing fetus and via breast milk. Although DDT levels are relatively low, fetal development represents a critical phase in neurological development and emphasizes the vulnerability of the fetus to neurotoxic agents such as DDT.

Neurodevelopmental defects have been associated with pre- and post-natal exposure to DDT. Effects include multiple neurological deficits, such as abnormal reflexes, impairments in memory, executive functions and social and attentional processes. Results from a 2021 study indicate that prenatal exposure to organochlorine pesticides was associated with higher risks of overweight in infants, with stronger effects among girls compared to boys. In addition, an investigation by Yin *et al.* (2021) demonstrated that prenatal exposure to 16 organochlorine pesticides, as determined in umbilical-cord blood samples, was associated with increased risk for neural tube defects in offspring.

Other organochlorine insecticides of relevance, including cyclodienes and hexachlorocyclohexanes, are differentiated from DDT not only on the basis of chemical structure but also on the mechanisms of neurotoxicity and defined ecological risks. Both groups include well-known insecticides extensively used in rural and residential environments.

The hexachlorocyclohexanes include a variety of isomers such as lindane and β-hexachlorocyclohexane. In several cases toxicity is equal to or greater than that of DDT, a feature contributing to the high efficacy of these insecticides for pest control. However, as seen in earlier examples, toxicity, coupled with persistence in the environment and the ability to accumulate, eventually culminated in the phasing out and banning of these organochlorines in the USA and elsewhere.

As with DDT, cyclodienes and hexachlorocyclohexanes primarily act on the central and peripheral nervous systems, inducing a sustained neuronal hyperexcitation that results in a rapid onset of convulsions and seizures. The aforementioned effects are attributed to the ability of these insecticides to inhibit ion transport by specific ATPases, in addition to interfering with the signalling of the primary inhibitory neurotransmitter, gamma-aminobutyric acid, in the central and peripheral nervous systems.

The neurological effects of the cyclodienes and hexachlorocyclohexanes centres on the association with specific neurodegenerative disorders such as PD. In particular, efforts have focused on the effects of cyclodiene exposure on disruption of the dopamine system implicated in the pathogenesis of PD. Other evidence points to the role of cyclodiene insecticides as risk factors for neurodevelopmental deficits, including autism spectrum disorder, but these findings await confirmation.

The characteristic nature of organochlorine insecticides, including cyclodienes, is the ease with which these compounds can cross the placenta to induce *in utero* exposure to the fetus and impact on neurodevelopmental processes. Although PD is perceived as an ageing condition, it is important to determine the potential role of prenatal exposure to organochlorine insecticides on the development and functioning of the dopamine pathways in the central nervous system.

Dieldrin is also associated with neurotoxicity, inducing rapid onset of convulsions following acute administration. In addition, chronic exposure to lower concentrations causes headaches, dizziness, muscle twitching and hyperirritability, most likely due to disruption of ionic homeostasis and gamma-aminobutyric acid signalling. However, the effects of dieldrin on the dopaminergic system and the possible association with the development of PD are worth noting.

Several epidemiological investigations indicate a link between levels of dieldrin found in post-mortem brain samples and PD. In particular, increased levels of dieldrin occurred in specific brain regions, including those uniquely damaged in this disease.

Similarly, other organochlorine insecticides including heptachlor and endosulfan have been associated with neurotoxicity and with disruption of dopaminergic pathways as the principal focus of attention. Of particular concern is the association with neurodevelopmental deficits in children following *in utero* exposure to endosulfan. Elevated levels of endosulfan have been found in cord blood and breast milk of mothers. There is added significance due to epidemiological evidence associating endosulfan exposure with PD and incidence of autism spectrum disorder.

Ongoing concerns are compounded by emerging evidence of toxicity in pesticides currently recommended for use by farmers. For example, the widely used soil fumigant, metam sodium has recently been associated with respiratory intoxications among farmers and close neighbours. In 2004, the US EPA designated this pesticide as a 'probable human carcinogen'. In 2018, regulatory authorities in France imposed a ban on products containing metam sodium, arguing that these pesticides constitute a risk to human health and ecosystems.

The inescapable conclusion is that organochlorine insecticides are neurotoxins. The association with the incidence, initiation and progression of neurodegenerative disorders is gaining momentum.

- Explain the role of 'sodium channels' in neurophysiology.
- What is the role of dopamine?

4.5.3. Organophosphates (OPs) are associated with both acute and chronic features of toxicity arising from widespread availability of different forms and uses of these compounds as insecticides. Emerging issues include:

- regular deployment as poisons;
- effects on the central nervous system;
- mitochondrial dysfunction;
- prenatal exposures;
- epigenetic modulation;
- potential risks in diabetes and hyperglycaemic states; and
- carcinogenesis.

The poisoning of five individuals in Salisbury (UK) brought into sharp focus the acute toxicity of OPs in a real-life situation. This assault in March 2018 caused the death of one person, with the four survivors requiring resuscitation and treatment in an intensive care unit. It has been confirmed that the OP used was Novichok, one of a class of highly potent neurotoxins, developed specifically as a lethal weapon. Anecdotal evidence indicated that the routes of exposure included skin absorption, inhalation and ingestion with food and drink. The expression of toxicity was progressive and extremely rapid, proceeding as follows:

- neurotoxicity;
- seizures;
- severe respiratory distress;
- coma; and
- death.

It is more usual to see chronic expressions of toxicity, as for example in farmers using OP-based sheep dips for controlling insect ectoparasites. Exposures occur via inhalation of contaminated aerosols and via direct skin contact.

OPs exert their toxic effects by irreversibly binding to the pivotal enzyme acetylcholine esterase. The failure of this esterase to hydrolyse acetyl choline results in an endogenous overflow at muscarinic and nicotinic synapses in the central nervous system and at the neuromuscular junctions in the peripheral nervous network. This process eventually leads to a cholinergic crisis, with:

- hypersecretion of glands (salivation, lacrimation);
- smooth muscle contraction (bronchoconstriction, urination, diarrhoea, abdominal cramps and emesis); and
- cardiovascular aberrations, including bradycardia, arrhythmia and hypotension.

Overstimulation of the perspiratory glands is mediated via the sympathetic nervous system and often observed in OP intoxication.

Nicotinic stimulation results in:

- muscle fasciculations;
- twitching;
- cramps;
- severe muscle dysfunction;
- respiratory failure; and
- death due to profuse secretions in the respiratory system and paralysis of the diaphragm and intercostal muscles.

Appearance of symptoms is dependent upon the dose and toxicity of the OP, route of exposure and metabolism within the body. First indications of toxicity occur at approximately 50% inhibition of the target enzyme, acetylcholine esterase, but life-threatening symptoms appear after over 80% inhibition of this enzyme.

Another proposed mechanism for the adverse effects of OPs is based on the induction of mitochondrial dysfunction, associated with reduction in activity of electron transport chain enzymes, ultimately affecting production of ATP.

In view of the foregoing evidence, it is particularly noteworthy that in 2020, the largest manufacturer of chlorpyrifos announced discontinuation of production of this pesticide for commercial reasons. However, preceding studies showed that low to moderate exposure to chlorpyrifos during pregnancy was associated with memory deficits and reduced IQ (intelligence quotient) in children. A new study (Chiu *et al.*, 2021) appears to extend this association of prenatal exposures to DNA methylation of peroxisome proliferator-activated receptor gamma and child development. It was argued that chlorpyrifos can traverse the placenta and affect fetal growth, neurodevelopment, cognitive skills and language performance.

Results of another 2021 investigation suggest that chlorpyrifos exposures may adversely impact systemic and immune cell phenotypes in hyperglycaemic states and diabetes-related organ damage.

Finally, exposure to OPs may cause cancer. Malathion, for example, can induce cancer-linked gene expression in human lymphocytes. According to the IARC, parathion and malathion are classified as 2B and 2A carcinogens, with the latter being associated with non-Hodgkin lymphoma and prostate cancer. The mechanistic evidence for malathion includes genotoxicity, oxidative stress, inflammation and receptor-mediated effects.

In summary, the diverse toxicology of OPs continues to cause concerns for human health. The need for further research is now clear, particularly in view of potential implications for biodiversity.

- How would you explain 'muscarinic'?

4.6 Fungicides

4.6.1. Select (c).

Fungicides commonly in use are predominantly organic in chemistry with diverse structural features, including a number of sulfur-containing compounds. The major classes of organic fungicides include:

- dithiocarbamates, for example:
  - maneb;
  - mancozeb;
  - thiram;
  - zineb;
  - ziram;
- benzimidazoles, for example:
  - benomyl;
  - carbendazim;
  - thiabendazole;
- dicarboxamides, for example:

- iprodione;
- vinclozolin;
- triazoles, for example:
  - propiconazole;
  - tebuconazole;
- anilinopyrimidines, for example:
  - cyprodinil;
  - pyrimethanil; and
- strobilurines, for example:
  - azoxystrobin.

Sulfur and copper salts are the predominant members of inorganic fungicides. For ecological reasons and also on the basis of efficacy, organic fungicides are the preferred compounds in current use. The site of action of compounds in the latter group varies with the chemical class of fungicides, with some acting by leaf contact and others after systemic transport to appropriate subcellular organelles. At the biochemical level, systemic fungicides act by diverse mechanisms including, for example, inhibition of ergosterol biosynthesis in fungal pathogens; safety is assumed on the basis that this pathway is absent in humans and other animals.

- Can you identify the sulfur-containing fungicides in the list above?
- What are 'systemic' fungicides?

4.6.2. The initial presumption that fungicides present low toxicological risks for humans was based on LD_{50} results with rats ranging from 800 mg kg^{-1} to in excess of 15,000 mg kg^{-1}, following oral intake. However, there are now profound concerns regarding chronic exposures to certain fungicides. As with several environmental contaminants, the association with neurodegenerative disorders is the predominant toxicological issue, particularly as genetic factors play a relatively minor role in the aetiology of such conditions.

Although significant progress has been achieved in the understanding of the pathophysiology of Parkinson's disease (PD), the environmental factors contributing to the initiation and progression of this disorder remain elusive. Nevertheless, it is important to consider emerging evidence relating to:

- PD;
- aminergic and glutamergic networks;
- neuronal excitability; and

- differential effects of:
  - ziram;
  - maneb;
  - azoxystrobin.

Residential and occupational exposure to ziram has been associated with markedly increased risk for the development of PD and for early-onset cases, including the display of the spectrum of both motor and non-motor symptoms. It is proposed that changes in neuronal excitability could represent an underlying factor in the increased risk for PD caused by exposure to ziram.

Maneb is another dithiocarbamate fungicide associated with neurodegeneration. Exposure to maneb in model systems resulted in modification of catalytic and allosteric protein thiols, which in turn caused mitochondrial dysregulation, ATP depletion and impaired metabolic adaptability. This cascade of effects resulted in neuronal apoptosis and neurodegeneration (Anderson *et al.*, 2021).

Using a zebrafish model system, it has been demonstrated that azoxystrobin exposure induces hepatic pathology, oxidative stress, lipid peroxidation and genotoxicity, with excess reactive oxygen species (ROS) implicated in the DNA damage. It is, therefore, possible that fungicides in other chemical classes may also be associated with adverse effects in humans, and non-target species in general, and may interact with different pesticides such as herbicides.

- Can you define 'apoptosis' and explain its significance?
- What is 'genotoxicity'?

4.7   Herbicides

4.7.1.   Select (c) and (d).

Herbicides in general are designed to act on biochemical processes that are unique to plants. The theory is that these compounds should not interfere with metabolic and physiological pathways in animal species, including humans. In terms of chemistry, some herbicides are organophosphates, while a few are organochlorines. Herbicides can be classified according to their chemical properties as shown below:

- phenoxyalkanoic acids:
  - 2,4-D;
  - MCPA;

- bipyridyls:
  - diquat;
  - paraquat;
- dinitroanilines:
  - dinitramine;
  - trifluralin;
- amino acid derivatives:
  - glyphosate;
- amides:
  - alachlor;
- ureas:
  - diuron;
- triazines:
  - atrazine;
  - simazine; and
- sulfonylureas:
  - chlorsulfuron.

More generally herbicides are classified on the basis of selectivity, mode of action and timing of use in the crop production cycle. Of particular relevance here is the use of non-selective herbicides which act on weeds as well as plants used by beneficial insect pollinators such as bees and hoverflies.

- Can you name other pollinators likely to be affected by pesticides?

4.7.2. The perception that herbicides are toxic to humans is based on comprehensive evidence obtained in work with mammalian models as well as epidemiological observations involving occupational exposures.

Two major developments serve to amplify continuing health concerns over the use of herbicides. In 2018, the US manufacturer of a glyphosate formulation was ordered to pay substantial damages to an operative who reportedly developed non-Hodgkin's lymphoma as a result of regular exposure to this herbicide. In 2020, an additional 52,500 US claimants alleged development of cancer due to similar exposures. The second development, also in 2020, related to the intended export of paraquat from the UK to other countries. Use of this herbicide is prohibited in the European Union (EU) on the basis of its toxicity.

Considerable evidence for toxicity has emerged in respect of particular properties of three herbicides:

- paraquat and its association with:
  - neurodegeneration;

- ○ dopaminergic dysregulation;
- ○ metabolic derangements;
- ○ interactions with maneb;
- glyphosate-based formulations, with differential effects on:
  - ○ gut microbes;
  - ○ amino acid incorporation into aberrant proteins;
  - ○ immunocompetence;
  - ○ carcinogenesis;
  - ○ neurotoxicity;
  - ○ Parkinsonism;
  - ○ toxicity, caused by the use of surfactants as adjuvants; and
- atrazine, and its relationship to:
  - ○ endocrine disruption.

Exposure to paraquat extends the hypothesis for an environmental dimension in the aetiology of Parkinson's disease (PD). Post-mortem PD brain samples are noted for astrocyte senescence, while cultured human astrocytes exposed to paraquat also become senescent. It is believed that paraquat and other environmental pollutants promote accumulation of senescent cells in the ageing brain which can result in dopaminergic neurodegeneration.

In different animal models, regular doses of paraquat can induce several of the pathological characteristics of PD, including loss of dopaminergic neurons in the nigrostriatal dopamine network. Other evidence indicates the increased incidence of PD in rural residents exposed to farm applications of paraquat and among workers using the herbicide without adequate protection.

Of immense concern are the synergistic effects associated with combinations of paraquat and other common pesticides, which in animal models replicate the full spectrum of clinical manifestations observed in PD patients. For example, paraquat with the fungicide maneb reduces dopamine synthesis and affects dopaminergic neurons as well as motor functions. Synergism is also evident among individuals living in close proximity to fields treated with this combination and, furthermore, risks for children are even greater than for adults.

Other neurodegenerative disorders, particularly Alzheimer's disease should also be considered in a multicomponent context, implicating paraquat and several insecticides operating additively or synergistically. Alzheimer's disease is a common progressive neurological condition characterized clinically as loss of neurons and associated functions. Although evidence for the aetiology of Alzheimer's disease

67

remains elusive, it is generally agreed that the major risk factors include age, genetics and environmental contaminants, particularly pesticide exposure. Epidemiological data imply higher incidence of Alzheimer's disease in rural locations compared to urban areas.

Most insecticides are designed to operate as neurotoxins against target invertebrate pests, so it is logical to implicate these compounds in the development of PD, Alzheimer's disease and other disorders of the central nervous system in humans. The supposed involvement of paraquat in neurodegenerative disease adds another dimension to pesticide toxicology.

Glyphosate is the active ingredient in the pervasive herbicide, Roundup®. Glyphosate is an analogue of the ubiquitous amino acid, glycine. It has been assumed that glyphosate would be relatively non-toxic to humans as its main action in weeds involves blockage in the shikimate pathway which is essential for plant survival and growth but absent in humans and other vertebrates. However, glyphosate is associated with:

- Disruption of the balance of gut microbes, favouring proliferation of pathogenic microorganisms, thus potentially resulting in inflammatory bowel disease and leaky gut syndrome.
- Neurotoxicity – at the metabolic level, it is also possible that glyphosate might excite certain receptors in neurons, potentially leading to neurotoxicity.
- Transport dynamics in the brain – in animal models, glyphosate is actively taken up by L-type amino acid transporters and translocated directly into the brain.
- Neurological impairments, including memory deficits, increased anxiety and motor functions.
- Structural antagonisms – it has been tentatively suggested that the adverse effects of glyphosate may reside in its structural analogy with glycine.
- Immunotoxicity – it has been proposed that aberrant proteins containing glyphosate may be recognized as foreign molecules by body defence systems, instigating an adverse immune response.

In addition, it should be noted that glyphosate was classified as 'probably carcinogenic to humans' by the IARC in 2015. This feature has been under further scrutiny in the US courts. It was concluded that the herbicide was the causal agent in the development of non-Hodgkin's lymphoma in an operative chronically exposed to the herbicide.

It should also be noted that glyphosate-based herbicides are more toxic to non-target species, including mammals, than glyphosate alone. It has now emerged that the surfactant used in glyphosate formulations may add to overall toxicity of the herbicide. It is clear that there is a need for a further evaluation of glyphosate safety, taking into account routes of exposure and nature of surfactants used in commercial formulations.

Atrazine is widely used despite regulations prohibiting its use in the EU due to toxicity. In a 2021 report, atrazine was characterized as an endocrine disruptor of the male reproductive system. Effects included:

- decreased testosterone production;
- reduced absolute weights of testis;
- compromised sperm quality with respect to:
  - counts;
  - motility;
  - morphological abnormalities; and
- histological damage to testes.

In further laboratory studies, prenatal exposure to atrazine was associated with cryptorchidism and hypospadias in F_1 male offspring (Tan *et al.*, 2021).

It can, therefore, be concluded that there is robust evidence that herbicides are harmful to humans. Associations with neurodegenerative disorders, carcinogenesis and endocrine disruption are particularly noteworthy.

- What is 'non-Hodgkin's lymphoma'?
- What are 'surfactants'?

4.8   Pesticide Resistance

4.8.1.   Insecticide resistance is a major issue in the control of target species such as mosquito vectors as well as insect pests of crop and horticultural plants. For example, *Aedes aegypti* and *Ae. albopictus* act as major vectors for the transmission of important arboviruses associated with human diseases in the tropics (Moyes *et al.*, 2017).

Chemical management is severely constrained by the emergence of insecticide resistance in these and other vectors. The main categories of neurotoxic agents used to control insect vectors include carbamates, organochlorines, organophosphates and pyrethroids. However, resistance to all four classes of insecticides has been detected in the

Americas, Africa and Asia, with regional variations in the degree of vector responses to the different compounds.

Multiple insecticide resistance mechanisms have also been reported in *Anopheles funestus*, a major vector of the malaria parasite. The geographical distribution of the mechanisms that have been demonstrated to be effective in the development of insecticide resistance is now emerging for both *Aedes* and *Anopheles* mosquito vectors.

Similarly, complex resistance mechanisms have been elucidated for field populations of common pests of agricultural and horticultural crop plants. Two studies exemplify the scale and severity of the problem. The tobacco cutworm (*Spodoptera litura*) is a significant polyphagous insect pest of vegetables, soybean, tomato, sweet potato, groundnut and cotton. Many field populations of this pest have developed high resistance to multiple insecticides, including organophosphates, carbamate, pyrethroids and novel-chemistry agents such as indoxacarb. Field populations of the aphid (*Aphis gossypii*) is another significant pest of cotton that is resistant to imidacloprid and thiamethoxan, among others.

A diverse range of mechanisms is associated with the acquisition of insecticide resistance, with particular reference to genetic, metabolic and even structural adaptations. Salient points include:

- regional variations;
- mutations;
- metabolic resistance:
  - role of insect glutathione transferases;
- complex expression profiles and regulatory processes:
  - specific insect-related interactions;
- upregulation in synthesis of physical structures; and
- symbiont-mediated mechanisms.

Mutations affecting vector proteins targeted by insecticides (referred to as 'target-site' mutations) are an important mechanism for the development of resistance. However, continued monitoring has been recommended for *Aedes* populations in order to ascertain any geographical variations in this mechanism (Moyes *et al.*, 2017).

In the case of insect resistance to pyrethroids and DDT, voltage-gated sodium channel mutations are considered to be a dominant mechanism in *Ae. aegypti* although there are geographical variations in the extent to which these mutations operate for the different insecticides.

The pre-eminent example of metabolic adaptation is reflected in the expression of genes for cytochrome P_{450} monooxygenases, glutathione S-transferases and carboxy/cholinesterases. In studies with the model insect *Drosophila melanogaster*, it was observed that resistance to DDT was associated with over-transcription of a single cytochrome P_{450} gene, identified as *Cyp6g1*, further confirming a genetic-metabolic component to insecticide resistance.

The results of other investigations indicate the metabolic role of insect glutathione transferases. These enzymes play a pivotal role in detoxification of endogenous as well as xenobiotic compounds, providing protection against oxidative stress. During this process, insecticides may undergo reductive dehydrochlorination or conjugation reactions with reduced glutathione. Upregulation of metabolic genes for glutathione transferases has been implicated in the *A. funestus* vector, specifically in respect of resistance to DDT and dieldrin.

Upregulation of physical mechanisms may also contribute to the development of insecticide resistance. For example, it is suggested that over-expression of cuticle protein genes may occur in the *A. funestus* vector and confer physical resistance to penetration of insecticide molecules.

In addition, there is evidence of a symbiont-mediated mechanism for insecticide resistance. Insects in the nymphal stages of development are able to acquire bacterial symbionts of the genus *Burkholderia* from the soil. Once in the gut, these bacteria are capable of degrading insecticides on behalf of the insect host.

It is thus clear that diverse mechanisms exist in insects to promote resistance to common and novel insecticides.

- Give an example of a 'conjugation' reaction.
- Can you define 'symbionts'?

4.8.2. A major concern among farmers and toxicologists alike is the widespread and rapid development of fungicide resistance in plant pathogenic fungi. The emergence of single target-site fungicides has facilitated the evolution of acquired and heritable reductions in the sensitivity of pathogenic fungi to these protective compounds (Massi *et al.*, 2021). Issues associated with the development of fungicide resistance include:

- mechanisms relating to:
  - mutations;
  - structural or other modifications to the target site that reduce binding to the fungicide;

- o enhanced production of target proteins;
- o induction or upregulation of alternative metabolic pathways to circumvent the original process inhibited by the fungicide;
- o metabolic disposal of the fungicide;
- o active transport or sequestration of the fungicide;
- o genetic variability of the fungal pathogen;
- o life cycle of the fungus;
- o reproduction rate of the fungus;
- o host diversity of the pathogen; and
- toxicological implications, particularly:
  - o proliferation of fungicides;
  - o increased frequency of fungicide applications;
  - o mycotoxin production.

Resistance to commonly used azole fungicides illustrates well the effects of multi-locus adaptations in phytopathogens. Resistance in these fungi is associated mainly with mutations of a gene encoding a protein in the ergosterol biosynthetic pathway. However, in some fungi, resistance can be acquired independently of the ergosterol pathway, via mutations in highly conserved genes encoding a vacuolar cation channel, a transcription activator and saccharopine dehydrogenase.

It is routinely recommended that fungicides that operate by the same mechanism should not be used simultaneously or consecutively to control plant pathogens in order to avoid selection of resistant populations of the fungus. It is salutary to note that *Plasmopara viticola*, the obligate parasite of grapevine species, shows resistance to almost all classes of fungicides. In China, attempts to overcome fungicide resistance in *Botrytis cinerea* resulted in the proliferation of over 470 registrations of fungicides. Another strategy often adopted is increased frequency of fungicide applications. The toxicological repercussions of these practices on biodiversity in agroecology are rarely considered.

Another concern over the development of fungicide resistance is the potential impact on mycotoxin production. Recent observations with *Botrytis*, *Penicillium* and *Aspergillus* pathogens serve to demonstrate the rising threat of fungicide resistance in plant pathogenic fungi and the risks for enhanced mycotoxin contamination in plant products.

It should be stressed that the relationship between fungicide resistance and mycotoxin production by phytopathogens remains to be established. Nevertheless, it has been apparent for some considerable time that new fungicides are devised on a regular basis and added to the growing mixture of POPs in the environment.

In conclusion, toxicological risks are associated with the proliferation of a seemingly endless repertoire of fungicides that may impact on human health and biodiversity.

- What is 'ergosterol'?

4.8.3. Modern herbicides are designed to target pathways that are unique to plants, for example photosynthesis. Adverse impacts on humans and wildlife should, therefore, be minimized. In practice, agricultural use of these herbicides is now increasingly perceived as a serious risk for human health and biodiversity.

In addition to toxicological issues, persistent and indiscriminate farm applications are associated with development of resistance to existing herbicides. Widespread and repeated uses of synthetic herbicides over the past 70 years have imposed powerful selection pressures on plants. It is consistently maintained that repeated use of xenobiotic chemicals has selected for a rapid evolution of resistance and not just in plants.

Of considerable concern is the emergence of herbicide-resistant species of weeds in arable ecosystems. Resistance is driving weed abundance on a global scale, with effects particularly noticeable in Europe and North America. A particular concern is that weed populations resistant to one herbicide are likely to exhibit resistance to multiple herbicide classes.

The development of herbicide-resistant crops, particularly soybean, has been instrumental in facilitating this evolution. It is estimated by Schütte *et al.* (2017) that intensive use of glyphosate-based formulations over the past 20 years generated at least 34 weed species resistant to this herbicide. The geographical spread of these species encompasses 37 countries. This development exposes agronomic flaws in the so-called glyphosate-'ready' strategy for soybean varieties. It has been concluded that excessive reliance on a single mechanism of action of herbicides tends to favour the emergence and dissemination of resistant weeds leading to biodiversity loss in agroecology.

The suggested molecular mechanisms involved in the development of herbicide resistance include:

- reduced absorption and translocation of herbicide;
- sequestration of the herbicide in the plant vacuole;
- metabolic degradation of herbicide;

- rapid cell death response;
- nucleotide polymorphism in the target site protein/enzyme;
- increased gene copies of target enzymes in response to herbicides;
- increased gene expression to enhance target proteins;
- interactions between target-site and non-target-site mechanisms leading to higher resistance levels; and
- extrachromosomal circular DNA-based amplification and transmission of herbicide resistance. It is postulated that circular DNA is linked to chromosomes and transferred to gametes during meiosis.

These divergent mechanisms impose diverse consequences for the spread, fitness and inheritance of resistance traits in crop and weed species exposed to repeated doses of herbicides.

Higher and repeated applications of herbicides or the development of new xenobiotic chemicals to overcome resistance will add to toxicological risks, particularly for wildlife species in agroecology and beyond.

- Can you define 'circular DNA'?
- Define the terms 'gametes' and 'meiosis'.

4.9 POPs as Endocrine Disruptors

4.9.1. Select (d).

Endocrine disruption is caused by a diverse range of organic compounds present as contaminants in the environment. Endocrine-disrupting chemicals include:

- PCBs;
- dioxins;
- pesticides;
- flame retardants;
- components of plastics;
- surfactants;
- stain-resistance coatings; and
- personal care products.

The US National Health and Nutrition Examination Survey has confirmed that the human population carries a significant burden of chemicals, including PCBs, DDT, polybrominated diphenyl ethers (PBDEs), phthalates, bisphenol A, personal care compounds and antimicrobials, with some existing as co-contaminants in the same individual (Lambertino et al., 2021).

Human health and reproduction are critically dependent upon a functional endocrine system. Hormones are secreted from glands/organs and transported in the bloodstream to act as chemical messengers, often at remote target sites, in order to coordinate and regulate physiological processes.

- Can you define 'endocrine'?

4.9.2. Endocrine disruptors affect functional physiology at different sites and systems in the human body (as well as other vertebrates). The role of specific organic pollutants in these effects is well documented. Salient issues include impacts on:

- onset of puberty, as affected by:
  - oestrogenic compounds;
  - pesticides;
- teratogenic risks;
- uterine functions;
- the placenta;
- the mammary gland;
- male reproduction and infertility;
- urinogenital system, as influenced by:
  - proximity to sources of specific contaminants;
- thyroid homeostasis;
- parathyroid tumours and halogenated contaminants: a clinical enigma;
- stress hormones; and
- insulin resistance.

Puberty is the stage during late childhood when secondary sexual characteristics emerge and reproductive capacity is attained. These changes are endocrine regulated and thus vulnerable to the effects of endocrine disruptors. Chemicals implicated include oestrogenic components of personal care products contributing to early development of puberty. It is generally acknowledged that puberty is occurring at an earlier age in girls today, particularly in the Western world and environmental contaminants such as POPs may be contributing to this phenomenon.

Peripubertal exposure to dioxins, furans and PCBs may disrupt timing of sexual development. A 2021 study of Russian boys residing near a plant that historically produced organochlorine pesticides indicated identification of seven exposure clusters, based on serum levels of dioxins, furans and PCBs (Plaku-Alakbarova *et al.*, 2021). Chemicals with similar number and positions of chlorine atoms clustered together. Shared persistence, metabolism or source of contamination may contribute to these groupings.

Another recent study has identified the molecular basis of PCB activity in oestrogen receptor activation (Wang *et al.*, 2021). The data confirmed that hydroxylated PCBs are the active forms implicated in endocrine disruption. The mechanism involves inhibition of a sulfotransferase which sulfonates oestrogen, a pivotal process regulating the cellular activity of this hormone.

Other evidence indicates that ziram can disrupt Leydig cell development during puberty, possibly by downregulation of the steroidogenic factor 1. The ovary is another critical organ as it not only produces steroid hormones but is also affected by xenobiotic endocrine-disrupting chemicals, for example parabens, methoxychlor and bisphenol A. Defects include anovulation, infertility, oestrogen deficiency, premature ovarian failure and ovarian cyst development, with some of these effects reproducible in animal models.

The uterus is a major hormone-sensitive organ and endocrine-disrupting compounds have been implicated in a number of disorders such as uterine fibroids and endometriosis. Elevated levels of some PCB congeners have been found in the abdominal fat of women with fibroids. Although there are genetic determinants, incidence of endometriosis may be linked to organochlorine pollutants.

An important focal point for the action of POPs as endocrine disruptors is the mammalian placenta, responsible for the production of progesterone and oestradiol for the maintenance of pregnancy. The enzyme responsible for the synthesis of oestradiol known as aromatase is inhibited by the fungicide ziram which combines with its steroid-binding site, thereby reducing production of oestradiol.

The mammary gland is also a potential site for the activity of endocrine disruptors as its development is under hormonal control. Benign breast abnormalities such as cysts and fibroadenomas have been attributed to endocrine disruptors present in underarm personal care products.

Male reproductive development is regulated by hormonal activity and is, therefore, affected by endocrine disruptors at all stages from the early embryo to adulthood. The androgen to oestrogen ratio is pivotal in male sexual development which implies that endocrine disruptors with oestrogenic or anti-androgenic activities are particularly critical.

Male infertility is recognized widely as a global issue, associated with several pathological and environmental factors. Results of a 2021 study in Pakistan indicated that serum organochlorine pollutants were higher in infertile men compared to controls, with few exceptions (Amir *et al.*, 2021). Serum DDT metabolites were significantly higher in infertile men.

Thyroid functions may also be affected by endocrine disruptors, taking into account the rising incidence of goitre even in regions where there is no natural deficiency of iodine, a critical component of thyroxine and triiodothyronine. There is evidence that exposures to PCBs, phthalates, brominated flame retardants and perfluorinated chemicals are associated with thyroid-disrupting effects.

Emerging evidence suggests a possible environmental dimension in the incidence and progression of hyperplastic parathyroid tumours in humans. This concern centres on the detection of PCBs, PBDEs and DDT derivatives and other insecticides in tumours. In particular, PCB-28 and PCB-49 congeners correlated positively with tumour mass.

Prenatal exposure to the pesticide lindane may predispose girls to subsequent development of metabolic syndromes by affecting insulin status. Separately, it has been established that lindane may impart insulin resistance in muscles by impairing hormonal signalling.

It is therefore clear that environmental contaminants in general, and POPs in particular, are consistently associated with endocrine-disrupting effects and that there is sufficient evidence to warrant action to safeguard human health and reproductive capacity.

• Why are oestrogens important?

4.10 Ecotoxicity

4.10.1. The impact of POPs on biodiversity is clearly evident in agricultural, arboreal and marine settings where adverse effects are maximized, creating existential risks for diverse species. Arguments inevitably focus on risks associated with organochlorine compounds in these ecosystems. Particular concerns are associated with, or relate to:

• rare plant communities;
• silvicultural practices;
• aquatic biodiversity;
• synergistic interactions as confounding factors;

- abundance of invertebrates;
- 'insect apocalypse';
- amphibians;
- small mammals;
- avian populations; and
- apex predators.

The original design of herbicides on the basis of targeting specific biochemical pathways that are unique to plants has engendered a false sense of security as regards potentially harmful impacts. Agricultural use of herbicides is now increasingly perceived as a serious risk to biodiversity, particularly in the light of widespread use of glyphosate-based formulations and paraquat. Adverse impacts have been reported in terms of both numbers and range of species.

The results of a study in western France reflects other observations indicating that herbicides are more effective in controlling rare plant communities than the more ubiquitous weed species. It is maintained that herbicide use has changed the profile of weed communities, with decreases in species sensitive to chemical agents.

Aerial applications of glyphosate are used in eradication of coca and poppy plants in Columbia. It is important to assess the risks for both human health and biodiversity in these ecosystems. Insight into likely implications for wildlife may be inferred from emerging evidence in silviculture. For example, in southern USA, herbicide use continues to increase in pine plantations associated with the timber industry. Following single applications of herbicide, impacts on biodiversity are minor and temporary, with a rapid recovery in floral diversity and wildlife habitat conditions. However, there is currently a shift in management practices to increase herbicide applications. Under these conditions, the prevailing opinion is that both plant diversity and wildlife habitat quality will be impaired.

Further ecological insight is provided by glyphosate treatment of managed conifer forest plantations in the USA. Glyphosate application increased turnover of bird species in treated plots. In addition, warbling vireos (songbirds that are deciduous specialists) declined in treated areas, suggesting that these species may be particularly sensitive to glyphosate exposure. Moreover, nesting success of certain species was significantly reduced by this herbicide.

In 2020, predictions of an insect 'apocalypse' were challenged as exaggerated due to limited geographical analysis, taxonomic constraints, biases in sampling and diversity metrics. However, considerable evidence is available to demonstrate severe risks for beneficial insects due to indiscriminate and excessive applications of pesticides. For example, chronic exposures to insecticides, including neonicotinoids, can alter the interaction between bumble bees and wild plants. Furthermore, herbicides may damage the floral diversity available to a variety of important insect pollinators required for crop production.

The contribution of glyphosate herbicides to declines in abundance of amphibians is under active consideration, but the evidence is still sparse, particularly for natural populations of these species. Surface waters near agricultural fields are popular habitats for amphibians, enabling feeding, reproduction and early-life stage development. These habitats are also likely to be contaminated with residues of herbicides and fungicides. It is expected that commercial formulations are likely to be more toxic than the active ingredients for many aquatic animals, including amphibians, due to direct skin absorption which may be facilitated by the inclusion of surfactants in the mixture.

Clay minerals have a much greater impact on paraquat adsorption than organic matter and this may account for the absence of effects on earthworms and other invertebrates in soil. In contrast, toxic effects have been observed in amphibians, fish and other aquatic species. Paraquat inhibition of testosterone production in amphibians has been attributed to the effects of oxidative stress caused by ROS generation. Fish are vulnerable to aquatic pollution as they are in direct contact with potential sources of pollution.

The toxicological risks for avian populations exposed to POPs is well documented. A 2020 study confirms that shell thickness of eggs of the common kestrel in the Canary Islands was negatively affected by DDE exposure. Thus, despite the total prohibition of DDT in Spain since 1986, background DDE levels remain elevated in rural settings.

The aquatic/marine ecosystem provides ideal conditions for bioaccumulation whereby a particular contaminant is concentrated successively up the trophic chain from prey to predator. Consequently penguins, seals, killer whales, polar bears and crocodiles become severely contaminated, thereby compromising growth, reproductive capacity and survival of the most vulnerable species. The risks are even greater with the bioaccumulation of two or more legacy POPs, for example PCBs and organochlorine insecticides.

In common with other predators, polar bears have long existed and, indeed thrived, at the top of the Arctic trophic hierarchy, relying almost exclusively on a diet of fatty ringed seals for vital nutrients. However, wild prey animals also provide polar bears with harmful and potentially dangerous levels of POPs such as PCBs and pesticides that bioaccumulate in sequential steps in the food chain. For example, the fatty tissues of seals are known to be burdened with a wide range of POPs and other potentially harmful contaminants. The relatively high concentrations of POPs in polar bears have been attributed to pollution from Europe, North America and Asia.

Despite the reliance on experimental observations, the overall weight of evidence points to biodiversity loss following exposure to POPs in different ecological settings. The existential risks for some wildlife populations are now indisputable, particularly when climate change is factored in to prediction models.

- What is 'silviculture'?

4.11 References

Amir, S., Tzatzarakis, M.., Mamoulakis, C., Bello, J.H., Shah Eqani, S.A.M.A. *et al.* (2021) Impact of organochlorine pollutants on semen parameters of infertile men in Pakistan. *Environmental Research* 195: 110832.

Anderson, C.C., Marentette, J.O., Prutton, K.M., Khatri, M., Reigan, P. and Roede, J.R. (2021) Maneb alters central carbon metabolism and thiol redox status in toxicant model of Parkinson's disease. *Free Radical Biology and Medicine* 162, 65–76.

Chiu, K.-C., Sisca, F., Ying, J.-H., Chen, P.-C. and Liu, C.-Y. (2021) Prenatal chlorpyrifos exposure in association with PPARγ H3K4me3 and DNA methylation levels in child development. *Environmental Pollution* 274: 116511. https://doi.org/10.1016/j.envpol.2021.116511

Golzadeh, N., Barst, B.D., Baker, J.M., Auger, J.C. and McKinney, M.A. (2021) Alkylated polycyclic aromatic hydrocarbons are the largest contributor to polycyclic aromatic compound concentrations in traditional foods of the Bigstone Cree Nation in Alberta, Canada. *Environmental Pollution* 275: 116625. https://doi.org/10.1016/j.envpol.2021.116625

Kallenborn, R., Hühnerfuss, H., Aboul-Enein, H.Y. and Ali, I. (2021) Chirality in environmental toxicity and fate assessments. In: *Chiral Environmental Pollutants*. Springer, Cham, Switzerland, pp. 279–305. https://doi.org/10.1007/978-3-030-62456-9_10

Lambertino, A., Persky, V., Freels, S., Anderson, H., Untermen, T. *et al.* (2021) Associations of PCBs, dioxins and furans with follicle-stimulating hormone and luteinizing hormone in postmenopausal women: National Health and Nutrition Examination Survey 1999–2002. *Chemosphere* 262: 128309.

Luo, T., Hang, J.G., Nakayama, S.F., Jung, C.-R., Ma, C.C. *et al.* (2021) Dioxins in breast milk of Chinese mothers: a survey 40 years after the e-waste recycling activities. *Science of the Total Environment* 758: 143627. https://doi.org/10.1016/j.scitotenv.2020.143627

Massi, F., Torriani, S.F.F., Borghi, L. and Toffolatti, S.L. (2021) Fungicide resistance evolution and detection in plant pathogens: *Plasmopara viticola* as a case study. *Microorganisms* 9(1), 119.

Moyes, C.L., Vontas, J., Martins, A.J., Ng, L.C., Koou, S.Y. *et al.* (2017) Contemporary status of insecticide resistance in the major *Aedes* vector of arboviruses infecting humans. *PLoS Neglected Tropical Diseases* 11(7): e0005625. https://doi.org/10.1371/journal.pntd.0005625

Plaku-Alakbarova, B., Sergeyev, O., Williams, P.L., Burns, J.S., Lee, M.M. *et al.* (2021) Peripubertal serum levels of dioxins, furans and PCBs in a cohort of Russian boys: can empirical grouping methods yield meaningful exposure variables? *Chemosphere* 275: 130027.

Quinete, N., Hauser-Davis, L.S., Lemos, L.S., Moura, J.F., Siciliano, S. and Gardinali, P.R. (2020) Occurrence and tissue distribution of organochlorinated compounds and polycyclic aromatic hydrocarbons in Magellanic penguins (*Spheniscus magellanicus*) from the southeastern coast of Brazil. *Science of the Total Environment* 749: 141473. https://doi.org/10.1016/j.scitotenv.2020.141473

Schütte, G., Eckerstorfer, M., Rastelli, V., Reichenbecher, W., Restrepo-Vassalli, S. *et al.* (2017) Herbicide resistance and biodiversity: agronomic and environmental aspects of genetically modified herbicide-resistant plants. *Environmental Sciences Europe* 29(1): 5. https://doi.org/10.1186/s12302 estrogen -016-0100-y

Tan, H., Wu, G., Wang, S., Lawless, J., Sinn, A. and Zhang, Z. (2021) Prenatal exposure to atrazine induces cryptorchidism and hypospadias in F1 male mouse offspring. *Birth Defects Research* 113, 469–484.

Wang, T., Cook, I. and Leyh, T.S. (2021) The molecular basis of OH-PCB estrogen receptor activation. *Journal of Biological Chemistry* 296: 100353. https://doi.org/10.1016/j.jbc.2021.100353

Weitekamp, C.A., Phillips, L.J., Carlson, L.M., DeLuca, N.M., Cohen Hubal, E.A. and Lehmann, G.M. (2021) A state-of-the-science review of polychlorinated biphenyl exposures at background levels: relative contributions of exposure routes. *Science of the Total Environment* 776: 145912. https://doi.org/10.1016/j.scitotenv.2021.145912

Yin, S., Sun, Y., Yu, J., Su, Z., Tong, M. *et al.* (2021) Prenatal exposure to organochlorine pesticides is associated with increased risk for neural tube defects. *Science of the Total Environment* 770: 145284. https://doi.org/10.1016/j.scitotenv.2021.145284

5 Fossil Fuel Pollutants

Questions

5.1 Overview

5.1.1. Compared to solar energy, fossil fuels are _____ and more _____.

5.1.2. The three major fossil fuels contributing to pollution include: (i) _____ _____; (ii) _____ _____ and _____; and (iii) _____.

5.1.3. 'The extraction, transport and processing of fossil fuels continue to cause severe pollution'. Do you agree that this statement:

    (a) is valid?

    (b) refers exclusively to historical issues?

    (c) is irrelevant in a book on environmental toxicology as human health is unaffected?

    (d) only impacts the profitability and reputation of fuel companies?

5.2 Crude Oil

5.2.1. The environmental and human health hazards associated with crude oil pollution are primarily due to contents of:

    (a) hydrocarbons

    (b) sulfur

    (c) polychlorinated biphenyls (PCBs)

    (d) other constituents

© J.P.F. D'Mello 2022. *Key Questions in Environmental Toxicology* (J.P.F. D'Mello) DOI: 10.1079/9781789248548.0005

5.2.2. Evaluate the impact of crude oil pollution on coastal habitats and marine biodiversity.

5.3 Shale Oil and Gas

5.3.1. The extraction of shale oil and gas by hydraulic fracturing is:

(a) safe

(b) also known as 'fracking'

(c) associated with human health and ecological risks

(d) an example of 'clean' energy source

5.3.2. Outline the potential impact of fracking technologies on bio-diversity in different ecosystems.

5.4 Coal

5.4.1. Evaluate the health issues for residents living near coal-fired power stations.

Answers

5.1 Overview

5.1.1. Insert: non-renewable; polluting.

There are finite limits to the global supply of fossil fuels. There is also unequivocal evidence of the polluting nature of conventional sources of energy provision. Attention is now turning towards solar, wind and tidal technologies.

• What are the limitations of renewable technologies?

5.1.2. Insert: (i) crude oil; (ii) shale oil [and] gas; (iii) coal.

A wide range of pollutants is associated with the extraction, transport and processing of fossil fuels. Utilization of these sources also results in the emission of a diverse array of harmful pollutants, as detailed in Chapter 3, this volume. Although high-profile incidents highlighted in case studies have attracted scientific investigation, it is important to note that pollution associated with these three energy sources occur regularly, and sometimes continuously, during normal operations.

- Can you identify the major company extracting crude oil in the River Niger delta?

5.1.3. Select (a).

The extraction, transport, processing and ultimate utilization of fuels are associated with severe pollution risks affecting human health, habitat ecology and biodiversity. The risks linked to fuel pollutants have emerged in specific case studies relating to the:

- *Torrey Canyon* oil spill in 1967 (UK);
- *Amoco Cadiz* oil discharge in 1978 (France);
- Ixtoc oil well explosion in 1979 (Mexico);
- *Castillo de Bellver* oil spill in 1983 (South Africa);
- *Piper Alpha* offshore oil and gas explosion in 1988 (UK);
- *Odyssey* oil spill in 1988 (Canada);
- *Exxon Valdez* oil spill in 1989 (Alaska);
- *Prestige* oil spill in 2002 (Spain);
- coal ash discharge at the Tennessee Valley Authority generation plant in 2008 (USA);
- *Deepwater Horizon* explosion in 2010 (Gulf of Mexico);
- *Wu Yi San* collision with an oil terminal in 2014 (Korea);
- wastewater release from shale oil and gas installation in 2017 (USA);
- major oil spill affecting nine coastal states in 2019 (Brazil); and
- oil discharge from wrecked tanker in 2020 (Mauritius).

The summary above demonstrates the worldwide and continuing risks associated with extraction and transport of energy resources. Severe risks of crude oil pollution continue unabated. For example, it has recently been estimated that 240,000 barrels of crude oil are discharged into the River Niger delta each year by an international energy conglomerate based in the UK.

Additional environmental risks are also associated with operations in shale oil and gas extraction (fracking) as well as coal mining. For example, the Environmental Protection Agency (EPA) identified almost 500 contamination incidents linked to fracking operations in the USA.

For particular locations, activities at gas and petrochemical plants often impact on the health and well-being of nearby residents due to diverse sources of pollution. For example, during 2019, residents near to gas flaring operations at a petrochemical plant in Fife (Scotland) complained about vibration, noise, light, odour and smoke pollution leading to health manifestations such as sleep deprivation, headaches, sore throat and asthma. In Louisiana (USA) a location has been identified as 'cancer alley', due to 50% higher malignancy rates attributed to pollution from petrochemical plants.

- Can you identify pollutants likely to cause malignancy in 'cancer alley'?

5.2 Crude Oil

5.2.1. Select (a), (b) and (d).

Crude oil and related downstream products contain a complex and diverse mixture of compounds, particularly:

- hydrocarbons including:
  - volatile organic compounds (VOCs);
  - alkanes;
  - polycyclic aromatic hydrocarbons (PAHs);
- sulfur compounds; and
- trace elements.

As a rule, all crude oils contain similar groups of hydrocarbons, but the proportions of individual compounds vary with the source of these fuels. The relative proportions and concentrations of hydrocarbons are altered rapidly on exposure in the environment after accidental discharge into oceans or soil. One class of chemicals of considerable environmental concern includes VOCs, particularly benzene, toluene, ethylbenzene and xylene. The volatility of short-chain and low-molecular-weight fractions facilitate evaporation in these ecosystems, but changes may also occur through dispersal, dissolution and microbial degradation.

The occurrence of PAHs contributes significantly to the acute and chronic toxicity of crude oil in the event of spillage into marine or soil ecosystems.

An important issue is the sulfur content of crude and refined oils used as fuel in vehicles and ships. The oxides of sulfur are hazardous for environmental acidification as well as for human health.

Depending on source, crude oil contains trace metals which may be transferred to seafood destined for human consumption. Arsenic, cadmium and mercury present particular concerns for coastal communities affected by drilling operations or accidental oil spills.

- Name other factors that contribute to environmental acidification.

85

5.2.2. Analysis of major incidents has provided evidence of profound detrimental impacts of crude oil pollution on fragile coastal habitats and marine biodiversity. Points to consider include:

- hydrocarbon profiles;
- weathering dynamics;
- soil and sediment contamination;
- coastal and marine ecology;
- use of dispersants;
- oiling;
- outcomes for:
  - fish populations;
  - cetaceans;
  - sea turtles;
  - bird populations;
- metal toxicity; and
- recovery issues.

Any analysis of effects of crude oil pollution will inevitably depend on the type and source of oil. Thus, the *Deepwater Horizon* spill differed markedly from other incidents in terms of scale, duration, source of emission (deep sea floor versus surface contamination) and management protocols applied during such events.

Significant changes occur during weathering of crude oil on surface slicks, as well as on contaminated sands, rocks and boulders. For example, following the *Deepwater Horizon* explosion, oxygen content increased in the hydrocarbon residues on these surfaces as a result of exposure to the elements. Oxygenated fractions constituted up to 50% of the mass of weathered samples, comprising hydroxyl and carbonyl functional groups in long-chain C_{10}–C_{32} carboxylic acids and alcohols. The uptake and metabolic/physiological impacts of oxyhydrocarbons in marine organisms require elucidation.

In assessments of sediment toxicity following the *Deepwater Horizon* accident, a triad approach involving determination of chemical contaminants, *in situ* biological effects and macrofauna community structure, indicated adverse effects as far as 25 km from the spill site. There were direct correlations between the presence of oil and biological and ecological effects of reduced macrofauna, abundance and diversity.

In the Gulf of Mexico, a considerable length of marsh shorelines was affected by the *Deepwater Horizon* accident, causing near-complete destruction of dominant species and therefore exposing the coastline to wave erosion. For example, populations of marsh periwinkles were suppressed in heavily oiled vegetation for over 5 years, with ongoing impacts and incomplete recovery.

Marine zooplankton exert important roles in determining ocean trophic cascades and modulating biogeochemical cycles. These micro-organisms are the principal prey for larval, juvenile and certain adult fish species. Crude oil pollution can impose both acute and chronic effects on marine zooplankton, depending on duration of exposure, type of oil, food availability, organism size and life-history stage. Data published in 2021, suggest that immediately after the 2019 oil spill off the coast of Brazil, benthic communities simplified in terms of reduction of species, richness and abundance, while opportunistic taxa increased. However, after 2 months, macrofaunal communities restructured, and effectively recovered to previous levels (Craveiro *et al.*, 2021).

Other researchers attribute the relative resilience of zooplankton communities to ecosystem connectivity, high fecundity, relatively short generation times and refugia in greater ocean depths. Nevertheless, factors such as the incidence of harmful algal blooms and the toxicity of dispersants and bacterial–phytoplankton interactions may determine eventual outcomes in zooplankton resilience to crude oil pollution.

Phytoplankton also contribute to, and modulate, interactions in marine ecology in response to crude oil pollution. For example, the formation and fate of aggregates is an important process that enabled sedimentation of oil on the seabed in the aftermath of the *Deepwater Horizon* accident. In coastal ecosystems, dispersed crude oil may also interact with suspended mineral particles to form oil-mineral aggre-gates, thereby creating conditions for oil biodegradation (Henry *et al.*, 2021).

Dispersants used in the clean-up operations are also toxic, imply-ing an additive effect in the biological and ecological impacts of crude oil pollution in marine and coastal ecosystems. Work with *in vitro* mod-els show that chemically dispersed oil is cytotoxic and genotoxic to sperm whale skin cells, possibly due to higher PAH uptake.

There are thus considerable and widespread risks for deep water species in general in the aftermath of oil pollution, including that in the Gulf of Mexico. For example, samples of fish species collected in 2010–2011 indicated a tenfold increase in PAH concentrations, with mean values above the recognized threshold levels for adverse biological effects. This contamination declined to baseline values by 2015/2016.

Photographic and aerial surveys as well as spatial analysis, in conjunction with chemical profiling of contaminated and stranded carcasses, indicate extensive oiling, affecting at least 16 cetacean species in the aftermath of the Gulf of Mexico spill. It is likely that adverse effects were due to inhalation, aspiration, ingestion and adherence to skin.

The impact of oil spills is, appropriately, evaluated in terms of mortality in populations of marine mammals and birds. Generally, however, relatively small numbers of carcasses are recovered. In the *Deepwater Horizon* spill, some reports indicated relatively minor environmental impacts due partly to low recoveries of carcasses. There are, nevertheless, consistent concerns that marine mammal mortality is underestimated in the wake of crude oil disasters.

Other evidence based on acoustic monitoring data also indicates that acute and chronic population-level impacts were underestimated. Comparisons with the *Exxon Valdez* accident are inevitable, despite differences in the sources and extent of pollution. In the year after this spill, 33% of the resident pod of killer whales and 41% of the transient group were lost, with neither group recovering to previous levels even two decades later.

Initial studies of common bottlenose dolphins following the *Deepwater Horizon* accident indicated evidence of hypoadrenocorticism, a manifestation of adrenal dysfunction previously observed in animal models exposed to crude oil. Furthermore, these dolphins developed moderate to severe lung disease including alveolar interstitial syndrome and pulmonary congestion. These symptoms are significantly greater in incidence and severity than those observed in dolphins not affected by oil pollution.

Endocrine disruption in bottlenose dolphins may well emerge as an additional effect of crude oil exposure. In the wake of the *Deepwater Horizon* accident, only 20% of pregnant dolphins delivered viable calves, compared to 83% in a reference population.

It should be stressed that the interaction between oil contaminants and ill-effects in dolphins is an evolving issue, but is likely to be multifactorial in aetiology. In a preliminary assessment, 69 persistant organic pollutants (POPs) including PCBs, polybrominated congeners and organochlorine pesticides were determined in blood and a subset of blubber samples at three Gulf of Mexico sites and an unimpacted reference location. It was concluded that the determined background levels of these POPs did not correlate with incidence of health abnormalities previously reported for coastal bottlenose dolphins exposed to oil pollution in the wake of the *Deepwater Horizon* explosion.

Sea turtles typify the risks for marine species exposed to crude oil due to their breeding and migration behaviour. These reptiles migrate to different habitats for reproduction and for foraging in the sea. Inhalation and swallowing contaminated water negatively affect growth and survival due to irritation of mucous membranes around the eyes, buccal cavity, lungs and alimentary canal. Absorbed PAHs can enter vital organs such as the lungs and liver, thereby adversely impacting on physiological functions. Following the *Deepwater Horizon* explosion, visibly oiled turtles in the Gulf of Mexico had higher concentrations of PAHs in tissues and biliary fluorescence PAH metabolites than unoiled animals.

As part of the natural resource damage assessment scheme, avian toxicity observations were instituted after the *Deepwater Horizon* explosion. The evidence indicates both physical and physiological deterioration following exposure to crude oil and its contaminants. External oiling affected flight patterns, implying that migration might be affected as well as increased heat loss and energetic requirements due to plumage damage. For example, heat loss of heavily oiled mallards and scaup can be significantly higher compared to normal values, affecting rehabilitation due to plumage deterioration and loss of water repellency, particularly in oiled scaup.

Other ecological threats are still emerging almost a decade after the *Deepwater Horizon* oil spill. For example, high concentrations of the genotoxic metals chromium and nickel have been observed in tissues, including skin biopsies, of whales in the Gulf of Mexico. Whether this is a legacy of the oil spill or associated with the clean-up operations remains unresolved.

The long-term recovery of populations of marine animals exposed to crude oil pollution is a matter of critical importance, reflecting efficacy of initial efforts at clean-up and rehabilitation and the impact of natural phenomena in dispersal of contaminants and microbial degradation of the constituent hydrocarbons. The *Exxon Valdez* oil spill into pristine northern waters in 1989 provides evidence of the extent to which recovery is possible. In the case of harbour seals, mortality was estimated at 300 in a population that had been declining and with 80% of seals contaminated with oil. However, these seals are now considered to have recovered, with population numbers stabilizing or increasing. Mortality in killer whales following the accident was also significant at 20% in 1990, compared with an expected level of 2% or less.

In contrast, sea otters, bald eagles, sea ducks and black oyster-catchers are now deemed to have recovered since the *Exxon Valdez* oil spill. Nevertheless, it should be noted that sea ducks, oystercatchers and other species feed in intertidal habitats and were, therefore, highly vulnerable to crude oil exposure in these zones. An estimated 15% of the pigeon guillemot population died from acute oiling and, in addition, an increase in nest predation of incubating adult birds and chicks contributed to mortality in this species.

In summary, crude oil pollution has caused severe and long-term effects on biodiversity in the aftermath of incidents such as the *Exxon Valdez* and *Deepwater Horizon* accidents. The effects on endangered species are of particular significance. Dispersants used in remediation add to the toxicity of crude oil pollutants for marine animals.

- Name a dispersant used in the *Deepwater Horizon* oil spill.

5.3 Shale Oil and Gas

5.3.1. Select (b) and (c).

The extraction of shale oil and gas involves directional drilling and hydraulic fracturing, a process also known as 'fracking'. Commercial application of this technology results in large volumes of waste-water containing a complex array of organic, inorganic and radioactive compounds with the potential to adversely affect human health and aquatic ecosystems.

Common organic substances appearing in wastewater include:

- viscosity modifiers;
- solvents;
- friction reducers;
- corrosion inhibitors;
- biocides;
- petroleum hydrocarbons;
- PAHs;
- organic acids;
- phenols; and
- heterocyclic compounds.

In the Marcellus Shale (USA), for example, wastewater exceeded relevant water quality standards for barium, strontium, copper, lead, chromium, mercury, zinc, cadmium, arsenic and benzene by as much as 1000-fold. Petroleum hydrocarbons derived from the shale may also contaminate wastewater. In addition, some VOCs may be released into the atmosphere as gases or vapour.

- Why are 'viscosity modifiers' used in fracking?

5.3.2. The extraction of shale oil and gas by fracturing technologies results in the production of wastewater with high salinity and content of diverse organic chemicals. This composition inevitably raises questions about potential impacts on biodiversity, particularly in aquatic ecosystems. Ongoing concerns are justified in the light of frequent water contamination incidents near wells and in the disposal of wastewater. Significant issues relate to:

- metal dynamics and bioaccumulation;
- toxicity in model organisms;
- morphological derangements; and
- developmental aberrations.

Elevated concentrations of mercury occur in streams near fracking sites leading to bioaccumulation and possibly biomagnification of this toxic element in invertebrates at different trophic levels. A riparian songbird has been used as a bioindicator of pollution caused by fracking on the basis that, as a predator feeding on macroinvertebrates, it may be exposed to chemical contaminants released into the ecosphere by fracking. Feather composition tends to reflect blood dynamics and toxicology during exposure to localized pollution. It was found that feather barium and strontium concentrations were significantly higher in birds originating from fracking sites compared to controls.

In studies with model aquatic organisms, including *Daphnia magna* and *Danio rerio* (zebrafish) among others, wastewater produced earlier in the fracking cycle was associated with higher toxicity compared to later samples. It is implied that although intrinsic salinity of the wastewater is a primary determinant of toxicity, dissolved organics significantly contribute to overall adverse effects for freshwater organisms.

Manifestations of deleterious effects in zebrafish exposed to fracking wastewater include:

- morphological deformities in embryos;
- decreased embryonic/larval metabolic rates; and
- altered expression of key cardiogenic and developmental genes.

This disruption in cardio-respirometry responses was attributed to organic components, overriding the influence of osmotic stressors in the wastewater.

Developmental abnormalities in zebrafish embryos exposed to sub-lethal concentrations of fracking wastewater may also be expressed in terms of decreased swim performance in juvenile stages of these species. Again, the organic fraction of wastewater has been implicated in reduced fitness and aerobic capacities of juvenile zebrafish.

It is clear, therefore, that there is a need for more detailed risk assessments to inform policy makers on ecological protection at fracking sites.

• Can you list the major organic components of fracking brines?
• What are the major chemicals contributing to salinity in fracking wastewater?

5.4 Coal

5.4.1. Residential proximity to coal-fired power installations is associated with potentially severe health risks, particularly for children and vulnerable adults. A number of factors should be considered in risk assessments, including:

• emission of particulates;
• coal ash residues;
• concentrations of toxic metals;
• radiation and uranium exposures;
• epidemiological evidence; and
• underlying health status.

Combustion of coal at these plants increases emissions of ambient particulates, principally $PM_{2.5}$. In addition, relatively large quantities of nanomaterials are also produced during this process. In particular, 2017 data indicate the incidental release of titania suboxides which may contaminate the local atmosphere.

Furthermore, the coal ash residue remaining after combustion presents particular human health and environmental hazards. This coal ash is stored in landfill sites and slurry ponds generally located close to residential areas often associated with socio-economic deprivation. Significant risks of contamination arise when these impoundments are damaged or not fit for purpose, leading to soil and surface water pollution.

Concentrations of metals in coal ash can be up to ten times that in the original coal, but other contaminants include arsenic, mercury, lead, cadmium, chromium, nickel, zinc and PAHs. High levels of these contaminants occurred in the largest coal ash discharge in 2008 at the Tennessee Valley Authority generation plant in the USA. Effluents from impoundments may contain major and trace mineral elements that exceed regulatory guidelines for drinking water and ecological safety.

In addition, coal ash is increasingly acknowledged as a source of radioactive elements and the risks are perceived by some observers to be greater than hazards established for certain nuclear reactors and comparable to those for nuclear waste. The radioactive contamination of coal ash is determined by the distribution and concentrations of the different isotopes in the original coal. Following combustion of coal in power plants, isotopes of uranium, thorium and ruthenium and associated decay products are partitioned into the gaseous products and the ash fraction.

Relatively high concentrations of radium isotopes have been found in coal ash, with radioactivity significantly exceeding levels occurring in ordinary soil. It is salutary to note, however, that even low levels of radiation are potentially harmful, with the propensity of isotopes to enter the body via the lungs and to circulate in the bloodstream, thus causing accretion in the bones and teeth.

Anecdotal and epidemiological observations indicate potentially severe public health risks exist for residents, particularly children and vulnerable adults, living near coal-fired power installations. Cumulative data over the past few decades indicate that all-cause and premature infant mortality were higher for communities located near coal-fired plants. In addition, increased incidence of respiratory disorders and emotional, behavioural and cognitive deficits occur in children from these high-risk areas compared to those living further away from coal-fired plants. The immunocompetence of children may also be compromised by exposure to particulates associated with combustion and coal ash.

The lung cancer risk associated with exposure to coal ash aerosols is an emerging issue. A variety of carcinogens occur in coal ash, including aluminosilicates, several toxic trace elements, nanoparticles and radionuclides. Silica, arsenic, cadmium and hexavalent chromium occurring in coal ash have all been linked to increased lung cancer risks. Ultrafine and nanoparticles are toxic to lung cells and are capable of inducing oxidative stress, cytotoxicity and genotoxicity. Of considerable interest is the potential reactivity of nanoparticles consisting of titania suboxides on lung alveolar membranes, with impacts that appear to be independent of other co-exposures.

Radioactive elements are concentrated in coal ash aerosols. Alpha particle-emitting radionuclides are considered to represent significant risk factors for lung cancer incidence in communities residing near coal-fired installations. There are also preliminary indications that DNA damage may be induced by coal combustion particulates by mechanisms that include oxidative stress.

Coal ash contains moderate to high concentrations of uranium due to enrichment from combustion of coal. Surveillance of residents living in a coal mining area of China indicated elevated uranium levels in urine and hair compared to reference values. It was implied that uranium exposures had occurred over a protracted period of time (Wufuer *et al.*, 2018).

In view of continuing worldwide use of coal as an energy source in domestic and industrial settings, there is clearly an ongoing need to safeguard the respiratory health of impacted communities. Aerosolized coal ash particulates have recently been implicated as a risk factor for neurodegenerative disorders, thus confirming the diverse range of effects of residential or occupational exposure to these and other contaminants.

- What are 'titania suboxides'?

5.5 References

Craveiro, N., de Almeida Alves, R.V., da Silva, J.M., Vasconcelos, E., de Almeida Alves-Junior, F. and Filho, J.S.R. (2021) Immediate effects of the 2019 oil spill on the macrobenthic fauna associated with macroalgae on the tropical coast of Brazil. *Marine Pollution Bulletin* 165: 112107.

Henry, I.A., Netzer, R., Davies, E. and Brakstad, O.G. (2021) The influences of phytoplankton species, mineral particles and concentrations of dispersed oil on the formation and fate of marine oil-related aggregates. *Science of the Total Environment* 752: 141786. https://doi.org/10.1016/j.scitotenv.2020.141786

Wufuer, R., Song, W., Zhang, D., Pan, X. and Gadd, G.M. (2018) A survey of uranium levels in urine and hair of people living in a coal mining area in China. *Journal of Environmental Radioactivity* 189, 168–174.

6 Metallic Elements

Questions
6.1 Overview
6.1.1. Heavy metals are _____ to living organisms and are excluded from the group of _____ mineral elements.

6.1.2. The four heavy metals of major significance in environmental toxicology include: _____, _____, _____ and _____; with the respective chemical symbols _____, _____, _____ and _____.

6.2 Mercury
6.2.1. Mercury is:

    (a) a rare earth element

    (b) essential for some aquatic organisms

    (c) indispensable for bone structure, in association with calcium and phosphorus

    (d) toxic to living organisms

6.2.2. Discuss the public health risks associated with exposures to mercury.

© J.P.F. D'Mello 2022. *Key Questions in Environmental Toxicology* (J.P.F. D'Mello)
DOI: 10.1079/9781789248548.0006

6.3 Lead

6.3.1. Lead is:

    (a) widely distributed in the environment

    (b) an integral component of vitamin C

    (c) well tolerated by humans due to adaptive responses

    (d) interacts with ultraviolet (UV) radiation in living organisms

6.3.2. Discuss the toxicology of lead in humans, including relevant comparisons with the adverse effects of mercury.

6.4 Cadmium

6.4.1. Cadmium is:

    (a) a metal with limited industrial properties

    (b) now a legacy contaminant, with negligible exposures in humans

    (c) a non-essential element

    (d) an element of continuing concern as a toxic pollutant

6.4.2. Write a short review on the toxicity of cadmium for humans.

6.5 Arsenic

6.5.1. Arsenic:

    (a) is of minor concern in environmental toxicology

    (b) is ubiquitous in different ecosystems

    (c) interacts with atmospheric ozone

    (d) is exclusively associated with instances of acute toxicity in humans

6.5.2. Review the toxicology of arsenic in humans, including evidence from different countries to emphasize salient features.

6.6 Nickel

6.6.1. 'Mine more nickel', urges the chief executive of a leading electric vehicle manufacturer in the USA. Prepare a case study to highlight environmental concerns with mining operations and recovery of nickel from electronic waste.

6.7   Lithium

6.7.1. What is the geographical context of the 'lithium triangle'? Discuss the environmental and human health impacts of lithium mining and recovery of this element from electronic waste.

6.8   Copper

6.8.1. In 2020, executives at an industrial conglomerate in Australia announced a decision to extend the life of a copper smelter refinery beyond 2022. Prepare a risk assessment relating to the toxicological implications of this declaration.

6.9   Manganese

6.9.1. According to Queiroz *et al.* (2021), manganese is 'the overlooked contaminant in the world's largest mine tailings dam collapse' which occurred in a Brazilian estuary. Also in 2021, a 'closed-loop' process was announced for recycling manganese and other potentially toxic metals from waste lithium-ion batteries of electric vehicles. Prepare a risk assessment for these two cases, incorporating other evidence to explain the toxicological ramifications of manganese as a pollutant.

6.10   Chromium

6.10.1. 'The intriguing case of chromium' appeared in the title of a research article published by Genchi *et al.* (2021). Prepare a case study to evaluate the strength of the evidence presented in the paper.

6.11   Electronic Waste

6.11.1. Summarize the toxicological implications of electronic waste recycling.

Answers

6.1   Overview

6.1.1.   Insert: toxic; essential.

Heavy metals are toxic to living organisms and classified in a group separate from the essential mineral elements. In addition, heavy metals are independent of any association with vitamins or other nutrients.

- Can you name an essential mineral element required by all living organisms?

6.1.2. Insert: mercury; lead; cadmium; arsenic; with the respective symbols: Hg; Pb; Cd; As.

These are conventionally classified as 'heavy metals', although this term is not universally accepted as an appropriate descriptor.

- Have you seen the film *Flint* relating to lead contamination?

6.2 Mercury

6.2.1. Select (d).

Mercury is unequivocally toxic to living organisms, with particularly robust evidence of adverse effects in humans. Risks arise due to a number of factors, including:

- chemical forms;
- biomagnification;
- food contamination;
- metabolic interactions;
- neurotoxicity;
- cardiovascular implications;
- intrapopulation variability;
- genetic dimension to toxicity;
- biomarkers; and
- role of selenium in pathophysiology of toxicity.

Mercury occurs in two toxic forms: (i) inorganic; and (ii) organic. Risks arise due to the ubiquitous distribution of these two forms, associated primarily with industrial activity and biological transformations.

- What is the chemical difference between the inorganic and organic forms of mercury?

6.2.2. Profound and diverse manifestations of toxicity have emerged following human exposures to mercury in well-documented poisoning incidents. Particular issues to emphasize include:

- systemic effects:
    o cellular dysfunction;
    o cardiovascular and pulmonary impacts;
    o digestive and renal disorders;
    o immunotoxicity;
    o central nervous system (CNS) deficits;

- legacy and ongoing issues for communities exposed to mercury in:
  - Iraq;
  - Japan;
- neurotoxicity of methylmercury:
  - risks relating to fish consumption;
  - binding to cysteine;
  - dynamics;
  - susceptible sub-groups:
    - abnormal pregnancy outcomes;
    - incidence of cerebral palsy;
- adverse effects of inorganic mercury;
- selenium interactions;
- mechanisms of toxicity;
- emerging research:
  - endocrine disruption; and
- continuing risks.

It is widely acknowledged that environmental mercury impacts profoundly on all major compartments and systems in the human body. At the cellular level, mercury alters membrane permeability due to its affinity for sulfhydryl and thiol groups. It is also associated with mito-chondrial dysfunction, oxidative stress and increased levels of reactive oxygen species (ROS).

In the cardiovascular and respiratory systems, mercury may con-tribute to chest pain, angina, bronchitis and pulmonary fibrosis. It can, in addition, inhibit digestive enzymes and increase incidence of inflammatory bowel disease, ulcers and renal cancer. The immune sys-tem may be impaired via deleterious effects on leukocytes and allergy responses.

Widespread incidence of poisoning occurred in Iraq during the period 1971–1972, involving the use of seed grain, treated with methyl-mercury fungicide. A total of 6000 hospital admissions were recorded, with over 400 deaths linked to this incident. Symptoms reported by affected individuals included paraesthesia (sensory impairment), ataxia, constriction of visual field (tunnel vision) and auditory deficits.

The pre-eminent case of methylmercury poisoning, however, relates to Minamata disease in Japan caused by industrial contamina-tion. Minamata disease is a debilitating methylmercury intoxication of the CNS, observed in Japan during the 1950s and mid-1970s and coinciding with an era of rapid economic growth. In both cases, vast quantities of methylmercury appeared in aquatic ecosystems as a result of industrial contamination. The coastal area of the Yatsushiro Sea including Minamata and the basin of the Agano River in Niigata were affected by this contamination. Inhabitants in these areas were exposed to methylmercury primarily via a diet of fish and associated processed products. Methylmercury is rapidly and very effectively absorbed from the gastrointestinal tract into the bloodstream and transported to all tissues of the human body.

Epidemiological observations indicated that the primary manifestations in the Minamata incident related to the development of neurological derangements, including intellectual deficits. This evidence supports physiological data generated over several decades. Methylmercury covalently binds to cysteine in the body to form cysteine-methylmercury, a structural analogue of the essential amino acid, methionine, which can traverse the blood–brain barrier and the placenta barrier via an amino acid transporter. Methylmercury is converted to inorganic mercury in the body and excreted. However, during turnover, methylmercury affects a variety of organs causing toxicity.

The most susceptible sub-group to mercury neurotoxicity comprises pregnant women and an extensive database has been compiled to follow the neurodevelopmental effects at very low exposures to methylmercury. The incidence of abnormal pregnancy outcomes, including stillbirth and spontaneous abortion increased significantly between 1956 and 1968 in Minamata and the vicinity. Male to female birth ratios also declined in the late 1950s. In addition, the incidence of cerebral palsy was markedly higher during the period 1955–1958 in fishing villages bordering Minamata Bay, ranging from 1% to 12%, compared to 0.2% for the general population of Japan.

It is known that inorganic mercury damages the kidney, liver, gastrointestinal tract and the cardiovascular system. Mercury exposure is also associated with modulation of plasma concentrations of thyroid hormones, implying an endocrine-disrupting effect. In addition, perturbation of reproductive processes by inorganic mercury results in fetal abnormalities.

Emerging data suggest that the preferential cellular target of mercury is selenium rather than the sulfhydryl groups of cysteine. It is claimed that mercury has an inferior affinity for thiol groups, whereas the bonding to selenium-containing entities is considerably stronger. Thus, although binding to thiol groups enables transport of mercury across membranes and facilitates tissue and excretion dynamics, it does not account for the oxidative stress and organ damage associated with its toxicity.

A comprehensive review of recent data indicates that the primary macromolecular targets of mercury are the selenoproteins of the thioredoxin system and glutathione peroxidase. It is suggested that mercury binds to the selenium component of these proteins, thereby inhibiting their functions and so irreversibly compromising intracellular redox status. Under these conditions, accumulating ROS precipitate a wide range of adverse effects on glutamate and calcium metabolism.

In a separate development, evidence emerging in 2021 indicates that the protective effects of selenium towards methylmercury toxicity may be weaker in fetuses compared to mothers, presenting enhanced risks for this sub-group.

In addition, interactions involving mercury and noradrenaline in the pathogenesis of essential hypertension and the metabolic syndrome in ageing populations are emerging (Pamphlett *et al.*, 2021). It is proposed that mercury accumulation in the adrenal medulla is age-dependent, increasing progressively until it is detected in 90% of autopsy samples from individuals over 80 years of age.

Finally, it is disturbing that some 60 years after unravelling the aetiology of Minamata disease, steps to curb environmental mercury contamination on a global scale are only now being considered. Meanwhile, mercury emissions from coal-fired power installations in China, India, Eastern Europe and parts of the USA continue unabated.

- What is the basic structure of 'cysteine'?

6.3   Lead

6.3.1.   Select (a).

Lead is a pervasive and toxic element, which has been subjected to extensive redistribution by anthropogenic activity. Mining, smelting and combustion have contributed significantly to lead pollution.

Human exposures occur via contamination of food and water. Lead poisonings on epidemic scales have been reported regularly over several decades, for example in 1738 in Devon (UK) through contaminated cider. In 2014, a major incident was reported in Michigan (USA) relating to lead contamination of domestic water supplies.

- What is 'smelting'?

6.3.2. The toxicology of lead is defined by diverse physiological manifestations amid concerns over continuing exposures in humans. An alert issued in 2021, based on work at Imperial College London (UK), indicated that lead from gasoline continues to contaminate air in London, 22 years after a ban on its use in petrol.

Salient issues include:

- risk sub-groups:
  - developmental exposures;
  - children;
  - internal dynamics;
- gut physiology and the microbiome;
- perturbation of cellular processes;
- epigenetic effects;
- lead-interacting proteins;
- neurological disorders affecting:
  - cognition;
  - behaviour;
- endocrine disruption;
- absence of threshold values;
- case report; and
- comparisons with mercury.

Absorbed lead is deposited in the bone, where it can persist with a half-life of 20 years. Fetuses, infants and children are considered to be at greatest risk following exposure to lead. *In utero* effects can occur when lead from the maternal blood crosses the placenta which may be compounded in the newborn receiving breastmilk. The fetal nervous system is more susceptible to lead as immature endothelial cells allow transport of the metal into the developing brain. Childhood exposure to lead can decrease brain volume, particularly in the prefrontal cortex in adults.

Lead uptake by the brain results in disruption of synapse formation in the cerebral cortex, while impeding development of primary neuro-chemicals, depressing neuronal growth and altering the integrity of ion channels.

Bone lead is generally more bioavailable during pregnancy and in children whose bones are developing. In children, bones can serve as a continuous source of internal lead exposure as their bones undergo regular restructuring. Children can also absorb lead more effectively than adults and the former also store more of the metal in soft tissues.

There is increasing recognition of the adverse effects of lead on the physiology and microbiome of the gastrointestinal tract (Liu *et al.*, 2021a). Lead can disrupt the gut barrier, including tight junctions, and increase permeability to inflammatory cytokines and microbial metabolites.

Chronic exposure to lead even at relatively low levels causes a wide range of effects, including irritability, attention deficits, cognitive disabilities, aggressive behaviours and autism spectrum disorder, particularly in children. It has recently been claimed that lead-related cognitive impairments are not widely appreciated by clinicians. Other effects include abdominal pain, headache, tremors, lethargy, depression, memory loss, lack of coordination, speech defects, numbness and tingling in the extremities, delirium, convulsions and even coma. These manifestations imply that lead impacts multiple organ systems, but the most pronounced effects are on the CNS and the peripheral nervous system. In adults, lead poisoning induces peripheral neuropathy, muscular weakness, fatigue and motor deficits. In children, the impact even at low exposures is greater on the CNS, leading to cognitive and behavioural disorders and possible interactions with respiratory disease.

There is limited evidence of later-life antisocial problems and criminal behaviour following prenatal exposure to lead. Furthermore, lead can damage nerves associated with sense organs and control of bodily functions which may result in later-life neurodegenerative/neurological disorders such as amyotrophic lateral sclerosis and Parkinson's disease, Alzheimer's disease and schizophrenia.

Lead is also associated with reproductive disorders, reducing fertility in males, while maternal exposure, for example in the final trimester of pregnancy, can increase the risks of spontaneous miscarriage, preterm birth and low birthweights in offspring. Bone lead accumulated after years of exposure can, in pregnant women, serve as a source of prenatal exposure due to the marked bone turnover normally associated with gestation.

Lead exposures during gestation and post-partum periods, at levels which are prevalent, for example, in the general US population, have been linked with permanent brain damage which may not be evident until later in the child's life in terms of cognitive deficits. The adverse impact of lead on brain function will depend on whether exposure occurred at the fetal or paediatric stage.

Other health risks with lead are emerging, including adverse effects on renal, cardiovascular and immune functions. Sufficient evidence exists for decreased renal function in adults with blood lead levels less than 5 µg dl^{-1}. Adult blood levels with less than 10 µg dl^{-1} are associated with hypertension and even all-cause higher mortality, as well as reduced immunocompetence.

Suggestions that there is no threshold range for lead toxicity in humans have provided the impetus for continuous monitoring of risks on a worldwide scale. For example, the frequent elevated concentrations of lead in particulate matter in Delhi (India) have been attributed to recycling of lead-acid batteries by small enterprises in the city.

The effects of childhood exposure to lead on pubertal development have been evaluated in a Mexico City population, while routine screening of blood lead levels in New York City residents resulted in a reduction of national threshold levels to 5 µg dl^{-1}. The need for surveillance is highlighted by the episode of childhood lead poisoning in Flint, Michigan (USA). This case resulted from a series of contributory factors, including corroding drinking water infrastructure, deficiencies in implementation of regulations by utility companies and inadequate government oversight of industrial operations.

Comparisons between the toxicity of lead and mercury are inescapable. Both metals are associated with neurological disorders, but the mechanisms may be different, although damage to the blood–brain barrier is a common feature. Disruption of tight junctions at the blood–brain barrier has been suggested for lead. In the case of methylmercury, blood–brain damage is attributed to upregulation of vascular endothelial growth factor expression, but only in the cerebellum. The blood–brain barrier of the cerebellum is considered to be vulnerable to mercury due to reduced expression of tight junction proteins. Another common feature is that both metals are recognized endocrine disruptors, prompting questions regarding interactions with organic pollutants such as polychlorinated biphenyls (PCBs) and various pesticides.

- Define 'cognition'.
- What are 'tight junctions'?

6.4 Cadmium

6.4.1. Select (c) and (d).

Cadmium is a metal with important and diverse industrial applications. It is a toxic non-essential transition group element, released into the environment as a result of both natural recycling and anthropogenic activities.

Anthropogenic sources include:

- mining;
- smelting;
- use of cadmium-containing fertilizers;
- combustion of fossil fuels;
- waste incineration; and
- landfill.

The principal global use of cadmium is with nickel in rechargeable batteries, but concerns over human health exposures have led to restrictions and ban of use in several applications.

- Can you identify a type of fertilizer that might contain cadmium?

6.4.2. A wide spectrum of adverse effects is associated with human exposures to cadmium (Genchi *et al.*, 2020). The results of case studies in Japan and elsewhere have contributed significantly to formulation of ongoing risk assessments. Current concerns relate to:

- kidney malfunction:
  - tubular reabsorption;
  - chronic disease;
- bone defects:
  - *itai-itai* disease;
- reproductive dysfunction;
- child health;
- carcinogenesis:
  - International Agency for Research in Cancer (IARC) classification;
  - breast cancer;
  - apoptosis in lung;
  - mortality;
- cardiovascular effects;
- gene expression signatures; and
- epigenetic modifications.

In the kidney, tubular damage is the critical dose-related effect of cadmium exposure, representing the primary adverse response. Long-term exposure impairs renal tubular reabsorption, reflected in increased levels of low-molecular-weight proteins in the urine. Chronic kidney disease and the development of end-stage renal disease are major disorders worldwide and there is evidence of an association between high cadmium exposure and mortality from renal diseases.

105

The high incidence of osteoporosis, a systemic skeletal disorder characterized by low bone mass and structural deterioration of bone matrix is a major public health issue, with resulting fractures contributing to reduced quality of life and life expectancy and increased maintenance costs for individuals and communities. *Itai-itai* disease, the most advanced manifestation of environmentally-induced poisoning, occurred in Japan after excessive and long-term intake of cadmium-contaminated rice. The disorder was characterized as a combination of osteomalacia, osteoporosis and renal damage, with affected individuals being prone to multiple fractures.

Other investigations demonstrate statistically significant associations between higher cadmium exposure and reduced bone mineral density in various populations and exposure levels, suggesting that cadmium exposure may be a contributory factor in the public health burden of osteoporosis.

Tobacco smoking is a probable confounding factor in cadmium-associated bone defects. Regular smoking increases cadmium exposure which correlates with blood and urine levels of the element. Nevertheless, there appears to be evidence of a link between low-level cadmium exposure and osteoporosis in the general population.

There is increasing evidence to indicate that *in utero* exposure to cadmium may correlate with negative pregnancy outcomes, particularly low birthweights of offspring and that this effect is restricted to girls. For example, in a birth cohort investigation in rural Bangladesh, urinary cadmium concentrations in early pregnancy were inversely associated with birthweight and head and chest circumferences in girls, while no such observations occurred in boys. There is also limited evidence that prenatal and childhood exposures to cadmium may adversely affect growth and that differences in attained weight and height are only apparent in girls.

Early-life cadmium exposure can be detrimental for the development of cognitive abilities in children. Limited evidence indicates that cadmium can cross the blood–brain barrier, while other mechanisms may involve the induction of ROS, disruption of calcium signalling or changes in neurotransmitters or epigenetic profiles.

In separate analyses, the IARC concluded that there is sufficient data to classify cadmium as a Group 1 carcinogen for humans. This evaluation was based on lung cancer risks following inhalation. It should be noted that the assessment of cancer risk was constrained by the limited number of cases associated with long-term high exposure, the inability to examine dose–response relationships across different investigations, and the difficulties in excluding confounding factors, particularly smoking.

In conjunction with nickel, cadmium may act in breast cancer initiation and progression by binding to specific cellular receptors and simulating the effects of oestrogen. The concept of 'metalloestrogens' has been advanced in this respect, although both cadmium and nickel are capable of binding to other cellular components, including a variety of tissue proteins.

Results of a 2017 study indicated that inhaled cadmium bioaccumulates in the lungs and transforms bronchial epithelial cells as well as penetrating to peripheral tissues. Chronic cadmium exposures are, therefore, associated with reduced pulmonary function, obstructive lung disease and bronchogenic carcinoma. Cardiovascular disease, including myocardial infarction and peripheral arterial disease may also occur.

Cancer mortality in residents of the cadmium-polluted Jinzu River Basin in Japan have been assessed in a 2018 study. Women with renal damage associated with high cadmium exposure were at risk of increased mortality due to malignant neoplasms, including renal and uterine cancers.

Modulation of the epigenome has been proposed as a key mechanism for the action of environmental stressors during development. Alterations in this process can impact health outcomes later in life. It has been suggested that cadmium exposure may affect DNA methylation. In a specific case study relating to children residing near a municipal waste incinerator in China, blood levels of cadmium correlated with global DNA hypermethylation. This evidence implies that cadmium is an environmental 'epigenotoxicant'.

It is clear that cadmium exposure is associated with mortality caused by renal disorders, cardiovascular disease and cancer. Furthermore, cadmium is increasingly linked to epigenetic changes that may reflect developmental toxicity of the element.

- Define 'osteoporosis'.

6.5 Arsenic

6.5.1. Select (b).

Concerns over arsenic relate to its ubiquitous occurrence and redistribution in the environment, particularly in water, natural sediments and soil. It is present naturally in diverse geological formations, for example in:

- arsenopyrite (FeAsS);
- lollingite (FeAs$_2$); and
- orpiment (As$_2$S$_3$).

Arsenic in these sources can be mobilized at pH 6.5–8.5, thereby contaminating groundwater and drinking supplies.

Anthropogenic sources include combustion of fossil fuels, smelting of iron ores and releases from the semiconductor industry and from historical gold-mine wastes.

- What are 'sediments'?

6.5.2. Chronic exposure to arsenic is a worldwide issue due to contamination of drinking water and food. It is estimated that 300 million people may be affected by arsenic poisoning in its diverse forms. Metabolism within the body ensures a wide spectrum of toxicological outcomes in humans. Manifestations include induction of:

- ROS:
  - interactions with selenium;
  - cellular signalling;
- cancer:
  - geographical disparities;
  - confounding factors;
- skin lesions;
- cardiovascular disorders;
- developmental abnormalities;
- neuropathy; and
- immunological impairments.

In common with selenium, arsenic induces cytotoxicity and genotoxicity through generation of ROS. The effects are more pronounced with methylated forms of the element compared to arsenite, resulting in more potent inhibition of enzymes. Methylated arsenic and its metabolites are well-established carcinogens, inducing toxicity by blocking pathways of selenium metabolism. The imbalance of selenium compounds may be associated with the generation of ROS which can impact on genomic stability and carcinogenesis. The arsenic–selenium interaction may also affect cellular signalling through activation of critical transcription factors (Ali *et al.*, 2021).

Inorganic arsenic is classified by IARC as a carcinogen, being linked with malignancies of the liver, skin, lungs and kidney. There is also increasing evidence implicating arsenic in the development of prostate cancer. Furthermore, recent research suggests that arsenic can disrupt stem cell dynamics during carcinogenesis.

It has been suggested that carcinogenic risks are greater for adults than for children and that oral exposure is more important than dermal contamination. Whether these differences are due in part to occupational or inhalation sources of exposures or whether cigarette smoking is a contributory factor in adults remains unresolved. Analysis of diverse ecological zones in Pakistan indicates that arsenic-laden dust is an important source, implying that inhalation is a significant route of exposure in certain cases.

Arsenic exposure is commonly accompanied by the development of skin lesions, but interpretation of data is confounded by differential susceptibilities within a population. Results of a 2021 case-control study in West Bengal (India), indicated that individual variations may be linked to depletion of S-adenosylmethionine, required for methylation of arsenic compounds in the body. In addition, epigenetic downregulation of a critical methyltransferase may be responsible for higher susceptibility to arsenic exposure among individuals in this study.

Other significant features of arsenic toxicity include disruption of cardiovascular, reproductive, nervous and immune systems, thus implying diverse mechanisms in the aetiology of these disorders. Chronic effects of relatively high arsenic levels in drinking water (up to 100 µg l^{-1}) are associated with peripheral arterial and coronary heart diseases. Evidence in the USA, China and Italy indicate that even lower levels, of about 50 µg l^{-1}, increase risks of coronary heart disease and stroke, but smoking may be a confounding factor.

In view of worldwide concerns over environmental endocrine disruptors, investigations continue on the effects of arsenic on reproduction in humans and other vertebrates. It is generally accepted that prenatal exposure to inorganic arsenic causes adverse gestation outcomes. Some epidemiological evidence indicates that arsenic can induce premature delivery, spontaneous abortion and stillbirths.

Exposure to arsenic is also associated with the incidence of neurological conditions, including arsenic-induced senescence in West Bengal and peripheral neuropathy throughout India. Epigenetic evidence indicates that senescence-linked microRNA is upregulated relative to unexposed controls, particularly in individuals with peripheral neuropathy.

Epidemiological and experimental evidence demonstrate that arsenic impairs immunocompetence, affecting both systemic and cell-mediated mechanisms. In particular, it modulates differentiation, activation and proliferation of macrophages, dendritic cells and T lymphocytes. There are concerns about the effects of arsenic during the sensitive periods of pregnancy and early life of infants, particularly regarding immune responses to natural infections and the efficacy of vaccines. There is also evidence that arsenic may prevent treatment of severe immune-related diseases.

- Explain the difference between systemic and cell-mediated mechanisms in immunology.

6.6 Nickel

6.6.1. Case study: 'Mine more nickel'

The mining of nickel and its use in batteries and as a catalyst raises the profile of toxicity in ecological, occupational and other settings. There are concerns over its bioaccumulation in fish obtained from contaminated surface waters. Risks may also occur at electronic waste processing plants to recover nickel and other high-value metals (Liu *et al.*, 2021b).

A study of nickel mining and metallurgical activities on the distribution of heavy metals in Levisa Bay (Cuba) revealed serious pollution of sediments, particularly nickel, iron, cobalt and manganese. High turnover of nickel in certain aquatic species may be related to enhanced efflux, suppressed uptake and compartmental sequestration.

The main toxic features of nickel include kidney and cardiovascular disorders, lung fibrosis and cancer of the respiratory tract, including a high incidence of nasal and lung cancer following occupational exposure. The IARC has classified soluble and insoluble forms of nickel as Group 1 carcinogens, while nickel alloys are considered to be possibly carcinogenic to humans. A high frequency of oral cancer incidence has been observed in Taiwan at sites with metal-contaminated soil or those in close proximity to metallurgical industries.

There are reports that nickel may act via genetic and/or epigenetic mechanisms. There is also limited evidence that nickel may replace iron in catalytic sites of certain critical enzymes and that toxicity and carcinogenicity are linked to these properties. In addition, nickel may bind to particular amino acids, including cysteine, histidine, glutamate and lysine in the active sites of other enzymes, thus inhibiting catalytic activity. Furthermore, nickel has been defined as a teratogen, causing congenital malformations following prenatal exposure in experimental models.

The CNS is another target organ sensitive to nickel toxicity, with accumulation in the brain causing oxidative stress and mitochondrial dysfunction. Allergy has also been reported when items containing nickel are in direct and prolonged contact with the skin.

- Can you name another teratogen?

6.7 Lithium

6.7.1. The 'lithium triangle' refers to Argentina, Bolivia and Chile as the countries with the largest deposits of lithium in the world. Other significant resources occur in Portugal, Spain and China. Mining is now operated on an industrial scale due to applications of the metal in the manufacture of lithium-ion batteries. Several case studies have emerged relating to environmental and human health impacts of activities at open-cast lithium mines. Risks are also associated with lithium in electronic waste recycling units.

Environmental concerns relating to lithium include:

- use of large quantities of water, a scarce resource in several of the countries cited above;
- generation and disposal of sludge and contaminated wastewater;
- emissions of gases causing air pollution;
- degradation of landscape and amenity loss; and
- adverse impacts on biodiversity in the vicinity of mines.

Lithium levels in surface and underground waters in the vicinity of electronic waste processing units can be higher than in the general environment. Plant uptake in contaminated soil is likely to be a significant pathway for lithium to enter the human body, particularly near unregulated electronic recycling units.

Regarding human health impacts, there are opposing factors to consider. For example, lithium is endowed with therapeutic properties (Eyre-Watt *et al.*, 2021). This evidence cannot be used to justify increased exposure to lithium for the general population, either via public water supplies or contaminated food.

In contrast, the results of several investigations point to a narrow therapeutic range for lithium, with severe toxicity even at 'normal' serum concentrations of the metal. Other evidence associates lithium toxicity with the incidence of a syndrome akin to Creutzfeldt-Jacob disease.

- Can you find out more about 'Creutzfeldt-Jacob disease'?

6.8 Copper

6.8.1. The proposal to extend the life of an Australian copper smelter refinery beyond 2022 should be considered in conjunction with another scheme to mine copper in a desert haven that US indigenous communities regard as consecrated ground. It is imperative, there-fore, to provide a comprehensive environmental risk assessment for copper mining as well as recovery from electronic waste.

Regarding epidemiological evidence, it is relevant to consider data published in 2007 relating to the effects of a copper smelter strike in the USA. It was previously observed that copper smelters were associated with approximately 90% of all sulfate emissions in the south-west states of the USA, including New Mexico, Arizona, Utah and Nevada. However, over the 8.5-month stoppage period, a regional improvement in atmos-pheric visibility coincided with about 60% reduction in concentrations of suspended sulfate particles.

The period of stoppage is clearly insufficient for any impacts on human health to appear in the local community. However, the investi-gators concluded that ambient sulfate particulate matter and associ-ated air pollutants might be linked to adverse human health outcomes. It was also predicted that reduction in this pollution would ameliorate likely mortality impacts for local communities. A 2021 study confirmed that copper smelting is a process with the greatest environmental impact, compared to other forms of electronic waste recycling.

Notwithstanding the preceding comments, it should be recalled that copper is an essential nutrient for humans and other living organ-isms. Thus, copper-containing enzymes that react with oxygen play pivotal roles in diverse biochemical processes. In addition, copper is a component of ceruloplasmin which is involved in promoting iron trans-port from cells via the iron-exporting protein ferroportin.

It is generally agreed that the interaction between environmental and genetic risk factors underlying Parkinson's disease requires eluci-dation. However, recent evidence confirmed that copper exposure is an environmental risk factor for this condition. *PARK2* patient neuropro-genitors show increased sensitivity to copper relative to manganese or methylmercury. Copper causes greater mitochondrial fragmentation and ROS production in *PARK2* mutant cell lines.

Recent data confirm the connection between neurodegeneration and copper exposure. For example, Patel and Aschner (2021) indicated that both copper and Alzheimer's disease target the hippocampus, cerebral cortex, cerebellum and brainstem, thereby affecting motor and cognitive skills. It was proposed that copper exacerbates neuro-toxicity by expediting plaque formation.

Copper pollution is a particular issue on pig farms due to its use as a growth promoter. Although basic requirements are relatively low at

3–6 mg kg^{-1} diet, young pigs increase growth rates in response to higher levels of supplementation, generally up to 170 mg copper sulfate kg^{-1} diet. This excess of copper is voided in the faeces, thereby contaminating soil and surface waters.

- What is the significance of 'plaque'?

6.9   Manganese

6.9.1.   Queiroz *et al.* (2021) highlighted the risks to estuarine biota and human health following this mine tailings dam collapse in Brazil. The rapid biogeochemical dynamics of manganese (and iron) within the impacted estuary were of particular concern. Elsewhere, recovery of manganese from electronic waste recycling is also perceived as the source of potential toxicity due to multiple routes of exposure.

Evidence from a variety of sources indicates the ongoing relevance of manganese toxicity affected by:

- fulfilling dietary requirements;
- inhalation;
- water contamination;
- occupational or residential exposures;
- neurobiological changes;
- cardiovascular effects;
- infant mortality; and
- regional issues.

It is important to recognize that manganese is an indispensable micronutrient for humans and other living organisms. Manganese is required for optimum functioning of several enzymes, including arginase I and II. However, excesses via contaminated air, water or food can result in toxicity. Manganese contamination of water is widespread and well documented, particularly near mining or battery recycling operations. Dietary exposure is exemplified by the observation that babies fed formula milk had higher manganese in hair compared to breast-fed infants.

An intriguing feature is the uptake of manganese directly from the nasal mucosa into the CNS via olfactory pathways, as demonstrated in animal models. Rapid appearance of manganese in the olfactory bulb 8–48 hours after exposure has been confirmed. The trigeminal nerve may also enable transfer of manganese from the nasal cavity to the brain.

The archetypical example of manganese toxicity is the development of symptoms that resemble Parkinson's disease in terms of postural deficits, bradykinesia, shuffling gait and speech impediments. The term 'manganism' is often used to denote these manifestations which may progress from initial psychiatric abnormalities.

Cardiovascular toxicity has also been associated with manganese exposure. Features include inhibition of myocardial contractions, dilated blood vessels and hypotension. In addition, underlying chronic liver disease may determine the fate of manganese in the body, causing excessive accumulation in the brain, since hepatobiliary excretion is the primary route for clearance of the metal.

Manganese in drinking water has been linked to increased infant mortality in regions where contamination is prevalent, as for example in Bangladesh and North Carolina (USA). Furthermore, higher manganese levels in hair correlate with increased hyperactive behaviours, impaired cognitive achievement and decrease in IQ of school-age children. Motor coordination and hand dexterity may also be adversely affected in children exposed to excess levels of manganese in soils bordering industrial metal-processing plants.

- Can you define 'bradykinesia'?

6.10 Chromium

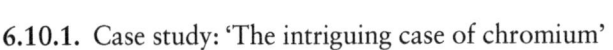

6.10.1. Case study: 'The intriguing case of chromium'

In reviewing the evidence for chromium, Genchi *et al.* (2021) considered wide-ranging factors to support this statement, focusing on:

- sources;
- prevalent oxidation states;
- functional metabolism in humans:
  - o contrasting properties;
- routes of exposure;
- comparative toxicology including:
  - o risk analysis;
  - o dermatitis;
  - o kidney disruption;

Continued

6.10.1. Continued.

o liver damage;
o mutagenicity and carcinogenicity;
o endocrine disruption;
o effects in plants; and
o destabilization of marine ecosystems.

Chromium is an industrial pollutant associated with manufacturing paints, corrosion-protecting agents and wood preservatives. Electronic waste is another important source of this element. It occurs primarily in two oxidation states, namely trivalent and hexavalent forms, with different properties. Trivalent chromium is an essential nutrient for humans.

In contrast, hexavalent chromium is a well-recognized toxicant released into the atmosphere as a result of industrial activity. It is classified as a hazardous air pollutant by the US Environmental Protection Agency (EPA). Inhalation of dusts and aerosols, skin contact and intake of contaminated food and water are important routes of exposure in different environmental settings.

Humans are normally exposed to hexavalent chromium in drinking water at levels of 0.2–2 µg l^{-1} due to natural erosion of soils and rocks or via contamination from industrial sources. The international limits for total chromium in drinking water range from 50 µg l^{-1} to 100 µg l^{-1}. Hexavalent chromium levels in groundwater in California (USA), Sao Paulo (Brazil), Northern Italy and Bangladesh range from 130 µg l^{-1} to 3000 µg l^{-1}, while the World Health Organization (WHO) limits for drinking water are set at 50 µg l^{-1}.

Hexavalent chromium is absorbed by cells via non-specific anion carrier proteins. Following entry, it is rapidly converted to the trivalent form, thereby yielding reactive chromium intermediates and ROS. These products modify cellular functions and facilitate apoptosis. Hexavalent chromium is a known irritant, readily penetrating the skin to cause inflammation and contact dermatitis. A marked effect of the element is its association with nephrotoxicity, with deposits in the kidney causing lesions in the proximal tubules.

Accumulation of hexavalent chromium may also occur in the liver, resulting in hepatotoxicity. It is suggested in a 2021 review that the liver is the specific target organ in its metabolic role of biotransformation and detoxification of xenobiotics, including hexavalent chromium.

Epidemiological and occupational evidence indicate that chronic exposures to hexavalent chromium are associated with the incidence of cancer. It is classified as a human carcinogen by the IARC, with

Continued

6.10.1. Continued.

links to prostate and lung malignancies. Cigarette smoking may compound the carcinogenicity of this form of chromium.

In addition, hexavalent chromium may act as an endocrine disruptor. The evidence relates to the marked prevalence of premature abortion and infertility in occupationally exposed women.

It is clear, therefore, that hexavalent chromium exerts multi-organ lesions in humans, while the trivalent form acts as an essential nutrient and that this is the basis for the intriguing case presented by Genchi *et al.* (2021). However, there are other metals that exist in different oxidation states that also serve as essential nutrients and cause toxicity. For example, elemental selenium is indispensable for humans but its tetravalent form as sodium selenite is highly toxic, as determined in a mammalian model. Again, elemental manganese is an essential element for all animals, including humans, but there is evidence that the oxidation state of this metal is an important determinant of tissue toxicodynamics and accompanying neurotoxicity.

- What is the chemical symbol for chromium?
- Define the term 'hexavalent'.

6.11 Electronic Waste

6.11.1. Electronic waste consists primarily of nickel, lithium, copper, manganese and chromium, but other elements such as lead, mercury and cobalt may also be present. Mining for these metals is often accompanied by environmental risks, including human morbidity and ecological degradation.

Electronic waste metals escape into the environment due to informal and unorthodox technologies employed in the manual dismantling, open combustion for metal recovery and illegal disposal of discarded fractions in sites exposed to inclement weather conditions.

It has been consistently reported that children living in the vicinity of electronic waste recycling sites endure health impairments due to toxic metal exposures. Health disorders include low birthweight, reduced anogenital distance, growth retardation, impaired pulmonary function and a greater prevalence of attention deficit/hyperactivity syndrome as well as higher DNA and chromosomal defects.

Immunocompetence may also be compromised in affected children. For example, in China, reduced antibody levels in preschool toddlers have been linked with lead in electronic waste, with almost 50% of chronically exposed children unable to develop adequate immunity to hepatitis in response to vaccination. It is implied that different immunization strategies may be needed for children living in conditions of chronic exposure to lead and, perhaps, other toxic metals.

In adults, occupational risks have recently been reported among workers at an informal electronic waste recycling unit in Ghana. These include musculoskeletal symptoms causing discomfort in the lower back, shoulders and knees. Disability was prevalent particularly in personnel operating as collectors and dismantlers (Acquah *et al.*, 2021).

- How do you dispose of your obsolete electronic devices?

6.12 References

Acquah, A.A., D'Souza, C., Martin, B.J., Arko-Mensah, J., Dwomoh, D. *et al.* (2021) Musculoskeletal disorder symptoms among workers at an informal electronic-waste recycling site in Agbogbloshie, Ghana. *International Journal of Environmental Research and Public Health* 18(4), 2055. https://doi.org/10.3390/ijerph18042055

Ali, W., Zhang, H., Junaid, M., Xu, N. and Chang, C. (2021) Insights into the mechanism of arsenic-selenium interaction and the associated toxicity in plants, animals and humans: a critical review. *Critical Reviews in Environmental Science and Technology* 51, 704–750.

Eyre-Watt, B., Mahendran, E. and Suetani, S. (2021) The association between lithium in drinking water and neuropsychiatric outcomes: a systematic review and meta-analysis from across 2678 regions containing 113 million people. *Australian and New Zealand Journal of Psychiatry* 55, 139–152.

Genchi, G., Sinicropi, M.S., Lauria, G., Carocci, A. and Catalano, A. (2020) The effects of cadmium toxicity. *International Journal of Environmental Research and Public Health* 17: 3782. https://doi.org/10.3390/ijerph.17113782

Genchi, G., Lauria, G., Catalano, A., Carocci, A. and Sinicropi, M.S. (2021) The double face of metals: the intriguing case of chromium. *Applied Sciences* 11(2): 638. https://doi.org/10.3390/app11020638

Liu, W., Feng, H., Zheng, S., Xu, S., Massey, I.Y. *et al.* (2021a) Pb toxicity on gut physiology and microbiota. *Frontiers in Physiology* 4 March. https://doi.org/10.3389/fphys.2021.574913

Liu, Y., Song, Q., Zhang, L. and Xu, Z. (2021b) Novel approach of *in-situ* nickel capture technology to recycle silver and palladium from waste nickel-rich multilayer ceramic capacitors. *Journal of Cleaner Production* 290: 125650. https://doi.org/10.1016/j.jclepro.2020.125650

Pamphlett, R., Jew, S.K., Doble, P.A. and Bishop, D.P. (2021) Mercury in the human adrenal medulla could contribute to increased plasma noradrenaline in aging. *Scientific Reports* 11, 2961. https://doi.org/10.1038/s41598-021-82483-y

Patel, R. and Aschner, M. (2021) Commonalities between copper neurotoxicity and Alzheimer's disease. *Toxics* 9(1), 4. https://doi.org/10.3390/toxics9010004

Queiroz, H.M., Ying, S.C., Abernathy, M., Barcellos, D., Gabriel, F.A. *et al.* (2021) Manganese: the overlooked contaminant in the world's largest mine tailings dam collapse. *Environment International* 146: 106284. https://doi.org/10.1016/j.envint.2020.106284

7 Consumerism and Lifestyle Choices: Toxicological Implications

Questions

7.1 Overview

7.1.1. The term '_____ pollutants' is still widely used to describe contaminants associated with consumerism and lifestyle choices despite universal agreement that some of these substances have been in existence for more than 50 years.

7.1.2. The three major categories of contaminants associated with consumerism and lifestyle choices, as classified in *Introduction to Environmental Toxicology* (D'Mello, 2020), include: (i) _____ and _____ _____; (ii) _____; and (iii) _____ _____ _____.

7.2 Plastics and Synthetic Fibres

7.2.1. Pollution caused by plastics and synthetic fibres is:

    (a) limited to the Pacific Ocean

    (b) restricted to European rivers

    (c) a global issue with extensive implications

    (d) confined to North American lakes

7.2.2. What are the human health concerns over the use and environmental distribution of plastics?

7.2.3. Discuss the adverse impacts of pollution caused by plastics and synthetic fibres on marine animals.

7.3   Pharmaceuticals

7.3.1.   Pharmaceuticals are:

(a) exclusively used in hospitals and clinics

(b) prophylactic and, therefore, devoid of harmful effects in humans

(c) subject to stringent regulations that prevent contamination of environmental compartments

(d) widely distributed in different ecosystems

7.3.2.   Briefly comment on 'antibiotic resistance' in bacteria.

7.4   Personal Care Products

7.4.1.   Personal care products are:

(a) major sources of polychlorinated biphenyls (PCBs)

(b) generally applied topically

(c) unlikely to damage fragile niches and ecosystems

(d) ubiquitous as contaminants in the environment

7.4.2.   Discuss emerging pollution and toxicological issues relating to the use of personal care products.

Answers

7.1   Overview

7.1.1.   Insert: emerging.
The term 'emerging pollutants' is widely used to denote:

- plastics;
- synthetic fibres;
- pharmaceuticals; and
- personal care products.

The supply-and-demand economics of recent decades has fuelled a relentless increase in the manufacture of these products. It is self-evident that these items have been contaminating the environment for some considerable time. Nevertheless, scientists continue to use 'emerging pollutants' as a collective descriptor for these contaminants (Dias *et al.*, 2021).

- Can you draw a map to illustrate the 'Great Pacific Garbage Patch'?
- Are you aware of local programmes to curtail waste associated with consumerism and lifestyle choices?

7.1.2. Insert: (i) plastics [and] synthetic fibres; (ii) pharmaceuticals; (iii) personal care products.

In a consumer-focused global economy, retail outlets strive to improve service to clients in terms of presentation, packaging and convenience. Meanwhile, society is preoccupied with the acquisition of goods irrespective of built-in obsolescence, resulting in an inexorable increase in pollution caused by discarded material. The insatiable desire to purchase articles with high turnover is a clear manifestation of prevailing throw-away attitudes, expressed vividly in affluent cultures.

In addition, the availability and increased use of pharmaceuticals and personal care products has compounded the risks for humans and species already made vulnerable by climate change and habitat degradation.

- What are the different categories of 'pharmaceuticals' that might pollute surface waters?

7.2 Plastics and Synthetic Fibres

7.2.1. Select (c).

Pollution caused by plastic products and synthetic fibres is widespread and not confined to European rivers or North American lakes. Extensive implications have emerged following media reports of visible and identifiable items discarded by retail outlets and consumers. Although ubiquitous in marine environments, rivers and estuaries, plastics are dispersed by rain, snow, prevailing winds and ocean currents to pollute pristine ecosystems around the world. This form of pollution is destined to continue due to increasing population, growing prosperity and rapid urbanization across the globe.

The assessment of pollution is complicated by the physicochemical diversity of debris classified as plastics and fibres. Items can range from plastic bags to minute particles. Over time, large objects and fragments will diminish in size due to photodegradation and physical abrasion to form microplastic particles.

Microplastics are defined as particles less than 5 mm in maximum length. This pollutant is either industrially manufactured as small-sized particles, as abrasive beads for cosmetic products or the result of physical breakdown (weathering) in the environment. A subset of ultrafine particles defined as nanoplastics are also of relevance, contributing special physiochemical properties.

Microplastics and nanoplastics are now ubiquitous, occurring in freshwater and marine ecosystems, including deep sea sediments. Regardless of origin, once plastic material enters the marine ecosystem either as large fragments or as microplastics, it will thereafter be degraded and dispersed within different ocean compartments, including the surface, the water column, sediment and living organisms. Plastic characteristics such as polymer type and density will also affect dispersal along these compartments.

In chemical terms, plastic polymers may comprise:

- polyethylene terephthalate;
- polystyrene;
- expanded polystyrene; and
- polypropylene.

The geographical distribution of plastics and synthetic fibres includes:

- the Pacific Basin;
- the Mediterranean Sea;
- the Greenland Sea;
- major waterways in Europe and Asia; and
- estuarine ecosystems.

The 'Great Pacific Garbage Patch', positioned between Hawaii and California, is the world's largest accumulation of ocean plastic. Within this region, specific interconnected areas have been identified as the 'subtropical convergence zone', the 'Eastern garbage patch' and the 'Western garbage patch'.

Also of considerable environmental concern is the 2019 headline that the Mediterranean Sea is 'dying' due to pollution with microplastics and other pollutants. Recent surveys indicate that the Greenland Sea represents a substantial reservoir of plastic particles.

In the UK, significant plastic pollution has been observed on the banks of the Thames estuary near Purfleet. It is estimated that a throughput of 94,000 microplastic particles per second flows downstream in the Thames. Contaminants varied from fibres in garments, microbeads in cosmetics to fragments of plastic packaging. Of equal concern is the evidence that all the major rivers in the UK are polluted with plastic debris, with the Mersey containing fragments, microbeads and fibres in concentrations exceeding those in the Great Pacific Garbage Patch. The Marine Conservation Society claim that almost two-thirds of shrimp in the North Sea contain synthetic microfibres which, ultimately, may be consumed by humans.

Rivers in Asia have been used for waste disposal for several decades, with plastic debris representing a large fraction of this pollution. The Ganges (India) is renowned for the complex nature of pollution, including plastics of varying dimensions. According to a 2019 survey, several waterways in Indonesia and Vietnam exceed pollution levels recorded in European rivers.

The environmental impact is exacerbated by reports that plastic waste from affluent countries in Europe is exported to Turkey for recycling, but is burned instead at the roadside. The emissions of styrene gas, polycyclic aromatic hydrocarbons (PAHs) and dioxins present particular hazards for the local inhabitants.

- What is 'styrene gas'?

7.2.2. Human health concerns, relating to the use and dispersal of plastics in different environmental compartments, are based on:

- biochemical and physiological impacts;
- food safety implications;
- transfer of adsorbed contaminants;
- spread of antibiotic resistance; and
- regular alerts.

Inferences from different investigations suggest a number of effects of ingested micro-/nanoplastics, including enhanced inflammatory reaction, size-related toxicity, transfer of adsorbed chemical pollutants and a dysfunctional gastrointestinal microbiome.

It is also suggested that these particles can be transported across living cells, such as the M cells in the gut mucosa or dendritic cells into the lymphatic and/or blood circulatory system to accumulate in secondary organs, including the liver and gall bladder, adversely impacting the immune system. It is further implied that nanoplastic size and hydrophobicity enable transport to pulmonary tissues and across placental and blood–brain barriers, potentially sensitive sites for toxicity. The relatively large surface-area-to-volume ratio in nanoplastics facilitates greater chemical reactivity, compared to microplastics. Similarly, it is proposed that relative to microplastics, nanoplastics may be more readily absorbed from the gut.

A major concern regarding microplastics as pollutants is their role as vectors of other contaminants, including constituent additives, toxic metals, organic compounds and pathogens. For example, results of a 2020 study indicated the propensity of plastic additives to disrupt oxidative metabolism and cause damage to macromolecules.

The influence of cosmetic microbeads on the adsorptive behaviour of cadmium and lead within intertidal sediments is of concern as this provides a route for metal contamination of freshwater and marine trophic webs, ultimately impacting on food safety for humans. Similarly, experimental work with polybrominated diphenyl ethers (PBDEs) adsorbed on to microbeads from personal care products demonstrate that ingested chemical pollutants can accumulate in fish, shellfish and filter feeders, thereby representing additional hazards for consumers.

In addition, questions have been raised over the role of microplastic particles in the transmission and persistence of antibiotic resistance in aquatic ecosystems. Adverse impacts on water safety and human health would be the inevitable result in such a scenario. Microplastic particles can accommodate specific microbial biofilm communities which can influence the overall microbiome within aquatic ecosystems by altering the balance of bacteria. It is known that these communities accumulate antibiotic-resistant genes; for example, sulfonamide-resistant bacteria commonly occur on microplastic particles. In a 2021 investigation, multidrug resistant pathogens, including *Vibrio cholerae* and *Citrobacter freundii* were isolated from plastic litter.

The pressure to continue investigative work has received impetus with recent public health alerts. For example, press reports in 2020 highlighted research indicating that bottle-fed babies swallow 'millions' of microplastics and nanoplastics each day. A 2021 report, based on data obtained with a rat model demonstrated the presence of plastic nanoparticles in maternal tissues, including placenta. Furthermore, these particles also appeared in fetal organs such as the brain, lungs, liver and kidney. Transfer across the blood–brain barrier in offspring was thereby implied.

- Comment on the significance of 'M cells'.

7.2.3. Adverse effects of plastics and synthetic fibres in marine animals are ascribed to physical and chemical factors, including:

- entanglement;
- ingestion and changes in feeding behaviour;
- structural characteristics of contaminating polymers:

- ○ microplastics;
- ○ nanoplastics;
- role as vectors of pathogens;
- chemical contaminants:
  - ○ intrinsic;
  - ○ extraneous;
- complex interactions with other pollutants;
- biochemical aberrations; and
- waste disposal constraints.

Entanglement with large plastic debris can cause starvation, suffocation, lacerations, infections and mortality in a wide range of marine and freshwater species. Access to, or competition for food and breeding activities may be hampered through entanglement with these items.

Limited evidence indicates that microplastics can cause behavioural changes, including, for example, reduced swimming activities of marine planktonic crustaceans and seabass, jump height of beach hoppers and predatory performance of common gobies, among others.

At the whole organism level, microplastics can affect feeding activity, body weight and energy reserves, reproduction success and even survival, with most of the data based on studies with invertebrates with short life cycles or with model species. For example, polystyrene nanoplastics can inhibit reproduction and induce abnormal embryonic development in the freshwater crustacean *Daphnia galeata*, a common model species used in ecological risk assessments. Transgenerational effects of microplastics in *Daphnia magna* have been observed, including parental mortality, reduced reproduction and population growth, leading to potential extinction of the population.

Of particular interest is the impact of microplastic pollution on the welfare and survival of marine animals. The relatively small size of microplastics confers biological and potentially adverse effects in a wide range of marine species at all trophic levels. The ingestion of microplastics, the common route of exposure, creates particular risks for organisms with indiscriminate feeding habits which tend to capture particles of similar dimensions to their natural food.

It is estimated that more than 200 species, including marine, freshwater and terrestrial organisms, have been observed to ingest microplastics. In some marine species, ingested particles can be retained and serve as obstructions in the alimentary canal or be translocated into tissues. Ingestion of microplastics by fish invariably leads to intestinal blockage, physical and histopathological changes in intestines, abnormal lipid metabolism and particle accumulation in the liver.

Microplastics are composed of a long chain of monomers with a mixture of additives, both attributed with toxic properties. Furthermore, following physical degradation the surface-area-to-volume ratio increases, enhancing the potential for adsorption, adherence and interactions with other non-polar toxic compounds and with microorganisms. Consequently, plastic fragments of diverse size can serve as vectors of pathogenic microorganisms, thereby creating risks for aquatic/marine species and the food chain.

Despite recent advances, there is still a need to correlate ecological end points with the physical characteristics of the diverse variety of plastics occurring as pollutants. It is known that fibres are more detrimental than particles for certain aquatic species and that irregular-shaped microplastics are more toxic than spherical beads for some crustaceans. In view of the abundance of synthetic fibres and irregular-shaped particles in different ecosystems, there is a need for further elucidation of some fundamental concepts in the toxicology of plastics.

In addition, composition of polymers may determine ecotoxicity of microplastics. Although plastics are considered to be almost inert due to their relatively large molecular size, reactions during polymerization are frequently incomplete, with monomers remaining within the polymeric materials being released during use and after disposal and degradation in the environment. Upon release, monomers can interact with cellular and molecular structures, leading to toxic effects. Some monomers, including those from polyvinyl chloride, are particularly hazardous to living organisms.

The relatively long chains of monomers, with strong chemical bonds resist biotransformation and degradation reactions, thereby conferring properties of persistence in the environment. Nevertheless, additives and other chemicals adsorbed on to plastic surfaces readily accumulate and may exert toxic effects or be neutralized by natural processes in oceans and sediments. Investigations with *D. magna* demonstrate that nanoplastics can interact in an additive mechanism with hydrophobic pollutants in aquatic ecosystems, underlining the high potential risks of these ultrafine particles.

It is estimated that plastics account for approximately 20% of global electronic waste production. These plastics are not recycled due to the presence of toxic substances, particularly heavy metals and brominated flame retardants. Laboratory tests, however, indicate that any leaching of such contaminants is unlikely to trigger a cytotoxic response (Shi *et al.*, 2021). While these results may be reassuring, the need for vigilance should be emphasized, given the established cytotoxicity of metal nanoparticles. Furthermore, experimental results show that plastic leachates consisting of PCBs and PAHs can induce severe developmental abnormalities in sea urchins.

In summary, there is compelling and visible evidence of harm caused by plastics and synthetic fibres in aquatic ecosystems. Type, physical characteristics and nature of adsorbed contaminants determine the ultimate adverse outcomes for marine species.

- Can you define 'monomers'?

7.3 Pharmaceuticals

7.3.1. Select (d).

Pharmaceuticals are subject to stringent regulations, particularly in developed and affluent countries. However, the manufacture and personal use of pharmaceuticals can result in significant wastewater discharges of active chemical compounds. As a result, diverse ecosystems may be impacted, affecting wildlife at various trophic levels and, in certain cases, ultimately creating risks for humans.

News alerts such as 'drugs in your drinking water', have brought into sharp focus the considerable limitations of current treatment and purification plants across the globe.

Pharmaceuticals of particular concern include:

- antibiotics, for example:
  - clarithromycin;
  - metronidazole;
  - amoxicillin;
  - 6-aminopenicillanic acid;
- recreational drug residues;
- psychotropic compounds;
- analgesics;
- anti-inflammatories;
- antihypertensives;
- illicit drug residues; and
- chemical (oral) contraceptives, for example:
  - 17α-ethinyl oestradiol.

Antibiotic pollution in rivers is widespread, even including the Thames (UK) and Danube in Continental Europe. For example, in the Danube clarithromycin was identified as an important contaminant, while rivers in Bangladesh contained excessive levels of metronidazole. In China, antibiotics occur in the major rivers, including the Yangtze and Pearl rivers and have also been detected in tap water. Antibiotics including amoxicillin and 6-aminopenicillanic acid as well as almost 70 others have been identified at high concentrations in tap/surface water. Pollution in soil and rivers occur via discharges in human and animal faeces, leaks from wastewater treatment sites and laboratories manufacturing pharmaceuticals.

A profound concern is the occurrence of pharmaceutical, recreational and psychotropic drug residues in surface waters in the Antarctic Peninsula, generally regarded as a pristine environment. Concentrations for 16 compounds ranged from nanograms to micrograms per litre (ng l^{-1} to μg l^{-1}), with maximum values for analgesics deemed to present particular environmental risks.

There is also accumulating evidence linking illicit drugs in wastewater to contamination of surface and drinking water in urban watersheds. It is clear that a number of these drugs are not effectively removed at drinking water treatment plants.

The synthetic compound, 17α-ethinyl oestradiol, is used in formulations of oral contraceptives. The active ingredient is excreted in the urine of women using this contraceptive and its worldwide occurrence at increasing concentrations contaminating major rivers and surface waters is yet another indicator of pharmaceutical pollution.

- Indicate the physiological classification of oestradiol.

7.3.2. Antibiotic pollution has long been recognized as a growing issue, raising concerns about the development of antibiotic resistance in bacterial species. The general disquiet is amplified by the indiscriminate and excessive use of:

- penicillin;
- tetracyclines;
- sulfonamides;
- β-lactams;
- vancomycin;
- virginiamycin; and
- an indeterminate number of other drugs with similar biological activities.

It is estimated that annual global deaths in human populations worldwide associated with antibiotic-resistant infections will increase to 10 million individuals by 2050 (He *et al.*, 2020). Despite substantial mitigation measures, incidence of clinical antibiotic resistance remains a persistent problem.

Over-prescription of, and continued reliance on, antibiotics in clinical settings has undoubtedly contributed to the emergence of resistant pathogens. The efficacy of a wide range of antibiotics has diminished as a result of the appearance of resistance genes. Even antibiotics considered as 'last resort' in certain medical applications are now associated with reduced effectiveness.

In addition, however, attention has focused on the impact of the use of antibiotics in animal feed for non-therapeutic purposes, and principally as growth promoters and, often, as a substitute for good hygiene on livestock farms. This practice has caused considerable disquiet in clinical settings, with arguments but no direct proof of the generation and proliferation of antibiotic-resistant strains of common bacterial pathogens.

It is maintained that feed antibiotics consistently appear within the gastrointestinal tract of livestock at low or sub-lethal concentrations. Under these conditions, growth of susceptible bacteria may be inhibited, but it causes selective pressure that ensures the development of antibiotic-resistant genes in surviving and other populations in the gut. Excretion of these bacteria in the faeces ultimately results in the spread of antibiotic-resistant strains into receiving environmental compartments such as soil and surface waters. Subsequent replication of resistant genes increases the risk of human exposure, directly or via manured food crops.

The occurrence of antibiotic-resistant genes in livestock waste generally corresponds to those classes that are regularly used as growth promoters. These include tetracyclines, sulfonamides and β-lactams, with the first two appearing in almost all livestock waste samples surveyed. As a general rule, the abundance of antibiotic-resistant genes in manure, farm wastewater and slurry is higher than that in background soils or upstream water sources.

Other factors should also be considered. For example, biofilms on plastic debris will accommodate bacteria with these genes and add to human exposures, for example during recreational use of rivers near livestock farms. Another route of dissemination of antibiotic-resistant genes/pathogens is transport via atmospheric particulates in polluted cities and regions across the globe.

It is worthwhile considering the specific issue of mobile resistance genes acquired via horizontal gene transfer. This process contributes significantly to the development of antibiotic resistance in populations of bacterial pathogens. Mobile resistance genes enable the spread of antibiotic resistance to new hosts, despite potential or actual taxonomic constraints (Ebmeyer *et al.*, 2021). Mobile genetic elements include:

- bacteriophages;
- conjugative plasmids; and
- integrating conjugative elements.

These elements are key mediators of genome evolution in bacteria, enabling the spread of resistance genes across bacterial populations. Transmission from donor to host bacteria occurs via conjugation, following cell-to-cell contact. Mobile genetic elements are, therefore, potent vectors for the spread of antibiotic resistance genes in bacterial populations even across taxonomic divisions.

In Asia and Africa, mobile genetic elements have exerted an important role in the spread of genes conferring resistance to combinations of antibiotics used, for example, against *Vibrio cholerae*. Mobile genetic elements can survive water treatment processes and spread to potential hosts, including *Escherichia coli*, *Mycobacterium* species, *Clostridium perfringens* and *Bacillus cereus*.

Processes concerning the development of antibiotic resistance in bacterial pathogens are also relevant. Three basic mechanisms have been proposed:

- deactivation;
- extrusion via efflux pumps; and
- protection of target sites.

An additional issue of concern is the 2020 conclusion that non-antibiotic pharmaceuticals can enhance the transmission of exogenous antibiotic resistance genes through bacterial uptake and transformation. Compounds such as anti-inflammatories and lipid-lowering drugs were implicated in this effect. It is relevant to enquire whether environmental contaminants, particularly persistent organic pollutants (POPs), might interact with mobile antibiotic resistomes distributed in soils and surface waters to enhance the. spread of antibiotic resistance in microbial pathogens.

It is important to emphasize that a clear link between antibiotic resistance genes in livestock faeces/slurry and the acquisition of such genes by the human gut microbiome currently remains unresolved. Nevertheless, it is prudent to reduce any risks by prohibiting the use of antibiotics in animal feed.

- What are 'bacteriophages'?
- Can you name the disease caused by *Vibrio cholerae*?

7.4. Personal Care Products

7.4.1. Select (b) and (d).

Personal care products are usually manufactured for topical use and classified in four major categories, as follows:

- sunscreens or ultraviolet (UV) filters;
- cosmetics;
- surfactants, classified as:
  - anionic;
  - cationic;
  - zwitterionic;
  - non-ionic; and
- insect repellents.

The chemical residues of personal care products are ubiquitous in the environment, with the potential to adversely impact human health and biodiversity. Although the above items are functionally different, similar environmental issues regularly arise due to use of common ingredients. For example, microbeads are routinely added to cosmetic products for cleansing and/or exfoliation of the skin. Microbeads are also regular components of sunscreens.

In addition, titanium dioxide is widely used in formulations of sunscreen, certain day creams, foundation and lip balms. Titanium dioxide is generally used in its nanoparticle form, serving as the primary inorganic UV filter.

Furthermore, the role of surfactants in optimizing the dispersion of nanoparticulate titanium dioxide-based UV filters in sunscreen for-mulations is under regular investigation in order to evaluate possible impacts on efficacy, environmental fate and toxicity. Of particular rele-vance are the physicochemical interactions between titanium nanopar-ticulate coatings and emulsifying agents containing surfactants.

Surfactants are widely used in other applications and, therefore, occur as pollutants from domestic, commercial and healthcare sources. Choice of surfactants may determine environmental impact, for exam-ple in freshwater and marine ecosystems.

Due to widespread transmission of mosquito-borne parasites and viruses, use of insect repellents has increased exponentially over several decades. Insect repellents formulated with the active ingredient, *N,N*-diethyl-meta-toluamide (DEET), are currently used by millions of people worldwide. It is credited with an 'excellent safety profile', providing high protection against mosquitoes, ticks and other potentially harmful arthropods. In other insect repellents picaridin, also known as icaridin, serves as the active ingredient.

- Which grade of sunscreen is recommended for children?
- Can you name a disease transmitted by an insect vector?

7.4.2. The emerging pollution and toxicological issues emanate principally from residues of the constituent active ingredients and additives included to enhance efficacy of personal care products. These compounds are not removed by standard wastewater treatment and regularly appear as pollutants in aquatic ecosystems. Efforts continue, nevertheless, to explore new methodologies for the extraction of personal care contaminants from wastewater (Li *et al.*, 2021).

It is instructive to consider likely impacts in the context of individual active ingredients and adjuvants, particularly:

- oxybenzone (benzophenone-3), including:
  - functions;
  - absorption;
  - properties:
    - mutagen;
    - carcinogen;
    - endocrine disruptor;
  - ecological impacts:
    - bioaccumulation;
    - coral-reef bleaching;
- octocrylene:
  - time-dependent degradation;
  - safety alert;
- titanium dioxide:
  - use as inorganic UV filter;
  - nanoparticle form;
  - safety concerns;
  - potential interactions with other environmental contaminants;
- microbeads;

- DEET;
- picaridin; and
- alert: perfluoroalkyl and polyfluoroalkyl substances in cosmetics.

Oxybenzone is an active ingredient of sunscreen lotions and similar products designed to protect humans against the carcinogenic effects of UV radiation (see Chapter 8, this volume). However, the dilemma between cancer protection and environmental damage should be considered. In addition, recent evidence indicates that 97% of people tested had oxybenzone present in the urine, presumably as a result of skin application and absorption. This compound is a photo-toxicant, implying that adverse effects are exacerbated by exposure to sunlight. Consequently, toxicity in humans is expressed as contact and photo-contact dermatitis, but there are also suggestions that oxybenzone may act as an endocrine disruptor and may be a factor to consider in current observations of reduced fertility in humans.

Oxybenzone is a contaminant of concern in marine/aquatic ecosystems. Treatment plants do not effectively remove oxybenzone or octinoxate (another UV filter) from effluents during standard processing procedures. Thus, current reports of various concentrations in waterways and fish are consistent with widespread contamination worldwide; the implications for bioaccumulation and seafood safety are, therefore, of some concern. Of particular relevance is that oxybenzone acts as a genotoxicant to coral planulae and cultured primary cells, producing effects such as reef bleaching and threatening the resilience of these reefs to adapt to other stressors.

According to Downs *et al.* (2021) and in view of definitive toxicological evidence, there is 'no safe harbor' for benzophenone in any personal care product under California Proposition 65 rules. This prohibition applies to UV filters, anti-ageing creams and moisturizers. However, concerns are also emerging over the substitute, octocrylene. Downs *et al.* (2021) claim that benzophenone accumulates over time from degradation of octocrylene in commercial sunscreen products. The authors urge that the safety of these products should now be 'expeditiously reviewed by regulatory agencies'.

Titanium dioxide is widely used in its nanoparticulate form as an inorganic UV filter in sunscreens, certain day creams, foundations and lip balms to provide protection against skin cancer development. However, concerns have emerged over safety of this filter. Its photoreactive mechanism can enhance reactive oxygen species (ROS) production, requiring coating of the nanoparticle with alumina or silica to suppress adverse impacts.

Furthermore, titanium dioxide nanoparticles can potentially penetrate the skin, lungs and gastrointestinal tissues and induce systemic toxicity. However, the overall lack of dermal absorption, as assessed by safety regulators, means that nano titanium dioxide is approved for use as an UV filter. Nevertheless, use in sprayable products is not permitted due to inhalation risks and lung exposure. This restriction is underlined by the International Agency for Research in Cancer (IARC) classification of nano titanium dioxide as a Group 2B carcinogen.

There are inevitable concerns that nano titanium dioxide, as used in rinse-off products, will contaminate aquatic ecosystems and accumulate in sediments of surface waters. Under these conditions, it is therefore possible that titanium dioxide nanoparticles will coexist and interact with other pollutants such as heavy metals and PCBs. The results of experimental studies with juvenile brown trout indicated that genes encoding for proteins and enzymes for tight junction functions and ROS elimination are significantly upregulated in the intestines of fish exposed to dietary combination of nano titanium dioxide and PCBs but not in single-substance treatments. A potentiating effect was, therefore, implied in this interaction.

Additional interactions may occur with the incorporation of surfactants and particle coatings in nano titanium dioxide products. It is argued that environmental risks strongly depend upon the concentration, aggregation state and surface characteristics of UV blockers. These factors may determine dispersal, environmental fate and toxicity of rinse-off products.

Surfactant pollution arises not only from domestic sources but also from facilities such as hospitals and commercial activities. It is argued that synthetic surfactants can enter animals through feeding or skin absorption. Furthermore, such exposures have been associated with toxicity due to the presence of synthetic surfactants in the blood, kidney, pancreas, gall bladder and liver of aquatic animals. In contrast, microbial biosurfactants are biodegradable, less toxic and more suitable for use in cosmetics.

Microbeads added to sunscreens and cosmetics represent a further dimension to the issue of plastic pollution. After use of rinse-off products, microbeads appear in wastewater treatment facilities and ultimately in rivers and marine habitats.

Regarding DEET, incidence of toxicity is rare and generally linked with incorrect or excessive use. Inhalational exposure, for example, can cause severe toxicity. Environmental contamination occurs via sewage treatment plants, with DEET being detected in groundwater below and adjacent to onsite wastewater treatment operations. In a 2018 study of pollution from Jakarta (Indonesia), DEET comprised a major proportion of personal care products discharged into the local marine ecosystem via a major river. Of particular concern is the first detection of DEET in sponges collected at two coral reef sites in the Maldives, attributed to sewage wastewater contamination.

With reference to aquatic toxicology, there are persistent demands for the need to undertake environmental risk assessments for insect repellents. DEET is moderately toxic to non-target insects such as the aquatic midge and caddisfly, by impairing feeding and development rates. This observation reflects recent data indicating that DEET is more toxic to some algae.

In contrast, high mortality in an aquatic predator of mosquito larvae was reported following experimental exposure to picaridin. It is implied that picaridin-based repellents in surface waters may increase the abundance of adult mosquitos due to reduced predatory pressure.

Media reports appearing in June 2021 alerted consumers to the widespread presence of 'forever chemicals' in major brands of cosmetics. The compounds of concern are perfluoroalkyl and polyfluoroalkyl substances used to increase durability and water resistance in products such as foundation, mascara and lipstick. Polyfluorinated chemicals are also used to make garments and carpets water and stain resistant, despite links to cancer and endocrine disruption. Ecotoxicity is, therefore, implied, impacting particularly on wildlife species.

It is now clear that use and disposal of personal care products increase the potential for pollution, adversely affecting human health and biodiversity. Particular environmental issues are associated with the release of active ingredients and functional adjuvants used in the manufacture of UV filters, cosmetics and insect repellents. The quest for 'natural' and biodegradable alternatives continues, but issues regarding efficacy and safety are inevitable.

- Have 'polyfluoroalkyl compounds' been mentioned in your course lectures?

7.5   References

Dias, R., Sousa, D., Bernardo, M., Matos, I., Fonseca, I. *et al.* (2021) Study of the potential of water treatment sludges in the removal of emerging pollutants. *Molecules* 26(4), 1010. https://doi.org/10.3390/molecules26041010

D'Mello, J.P.F. (2020) Consumerism and lifestyle choices: toxicological perspectives. In: *Introduction to Environmental Toxicology*. CAB International, Wallingford, UK, pp. 97–105.

Downs, C.A., DiNardo, J.C. and Lebaron, P. (2021) Benzophenone accumulates over time from degradation of octocrylene in commercial sunscreen products. *Chemical Research in Toxicology* 34, 1046–1054.

Ebmeyer, S., Kristiansson, E. and Larsson, D.G. (2021) A framework for identifying the recent origins of mobile antibiotic resistance genes. *Communications Biology* 4, 8. https://doi.org/10.1038/s42003-020-01545-5

He, Y., Yuan, Q., Mathieu, J., Stadler, L., Senehi, N. *et al.* (2020) Antibiotic resistance genes from livestock waste: occurrence, dissemination, and treatment. *npj Clean Water* 3, 4. https://doi.org/10.1038/s41545-020-0051-0

Li, Y., Zhang, C. and Hu, Z. (2021) Selective removal of pharmaceuticals and personal care products from water by titanium incorporated hierarchical diatoms in the presence of natural organic matter. *Water Research* 189: 116628. https://doi.org/10.1016/j.watres.2020.116628

Shi, P., Wan, Y., Grandjean, A., Lee, J.M. and Tay, C.Y. (2021) Clarifying the *in-situ* cytotoxic potential of electronic waste plastics. *Chemosphere* 269: 128719. https://doi.org/10.1016/j.chemosphere.2020.128719

8 Radiation Hazards

Questions

8.1 Overview

8.1.1. As listed in *Introduction to Environmental Toxicology* (D'Mello, 2020), radiation hazards arise from: _____ _____, _____ and _____ _____.

8.2 Ionizing Radiation

8.2.1. Apart from legacy issues caused by warfare, the sites of three major accidents remain as significant sources of radionuclides. In chronological order, these events occurred at: _____ _____ _____, _____ and _____.

8.2.2. Discuss the public health impact of ionizing radiation, using evidence from case studies to support your arguments.

8.3 Radon

8.3.1. Radon is:

- (a) a radioactive gas
- (b) a synonym for argon
- (c) associated with stratospheric ozone in its environmental distribution
- (d) affected by greenhouse gases in its activity

8.3.2. Outline the human health effects of radon exposure, including reference to any confounding factors that might influence final outcomes.

8.4 Ultraviolet Radiation

8.4.1. Ultraviolet (UV) radiation:

(a) comprises three types of radiation based on electrophysical properties

(b) interacts exclusively with volatile organic compounds (VOCs) in the troposphere

(c) is absorbed by the skin of humans

(d) is required for vitamin A synthesis in the liver

8.4.2. Discuss the effects of UV radiation on human health and morbidity.

Answers

8.1 Overview

8.1.1. Insert: ionizing radiation; radon; ultraviolet radiation.

Radiation hazards remain relevant, with worldwide implications despite considerable controversy. Evidence concerning the wide-ranging effects of ionizing radiation is based primarily on an abundance of historical case studies.

It is also becoming clear that the combustion of coal and shale oil and gas extraction by fracking are associated with low-level radioactive emissions.

Additional human health risks arise as a result of exposure to radon in dwellings and to ultraviolet (UV) in sunlight. Momentum has been stimulated by enhanced surveillance data for radon distribution and elucidation of the role of UV–vitamin D interactions in different manifestations of human morbidity (Moghadam et al., 2021).

- Have there been any concerns over the disposal of radioactive waste in your country?

8.2 Ionizing Radiation

8.2.1. Insert: Three Mile Island; Chernobyl; Fukushima.

These accidents occurred in the USA in 1979, Ukraine in 1986 and in Japan in 2011, respectively. Exclusion zones established at the three sites for safety reasons remain to the present day. However, regular radiation discharges from nuclear power plants and storage facilities on both sides of the Atlantic have highlighted the need for continued vigilance and monitoring of local communities for deleterious effects. Momentum for this work is destined to continue, fuelled by several alerts issued recently. Examples include:

- nuclear reactor still smouldering at Chernobyl;
- radioactive hotspot in the Chernobyl exclusion zone;
- installation of a new nuclear waste repository in the vicinity of the failed nuclear reactor at Chernobyl;
- higher than usual levels of ruthenium and caesium isotopes detected in Estonia, Finland and Sweden attributed to release from a nuclear power reactor in Russia;
- recommended return of residents to Fukushima;
- release of tritium-contaminated water from the failed Fukushima nuclear reactor into the Pacific Ocean;
- commissioning of a floating nuclear installation in Russia;
- high radioactivity in the Marshall Islands;
- fire at Iran's Natanz nuclear plant, attributed to cyber sabotage;
- commissioning of a new nuclear power station in Belarus, despite unresolved safety issues;
- 'serious' incident at a Finnish nuclear reactor, resulting in a radiation spike;
- concern over radiation risks at a nuclear plant in China; and
- disposal of toxic waste from the nuclear industry: significant and unresolved issues.

- What is 'tritium'?

8.2.2. The public health impact of ionizing radiation depends upon a number of critical factors including:

- source and distribution:
  - Chernobyl;
  - Fukushima;
- nature of emissions, particularly:
  - radioiodine;
  - radiocaesium;
  - strontium;
  - plutonium;
  - isotope decay;

139

- pathways of entry into the body;
- time-course dynamics;
- dose–response effects:
  ○ acute toxicity;
  ○ epidemiological observations;
- increased incidence of malignancy:
  ○ thyroid carcinoma:
    – developmental exposure;
    – age effects;
  ○ leukaemia;
  ○ breast cancer;
  ○ confounding issues;
- legacy issues, including:
  ○ the Radiation Effects Research Foundation's Life Span Study (Japan); and
- contingency and advisory measures.

Considerable toxicological evidence has accumulated since the detonation of two atomic bombs over Japan by the USA and major accidents at nuclear power generation installations in the USA, Ukraine and Japan. The explosion at the Chernobyl nuclear reactor caused widespread and long-term contamination of the environment, adversely impacting air quality, food safety and human health. This accident resulted in the release of volatile radioactive elements, including radioiodine (^{131}I) and radiocaesium (^{137}Cs), across extensive areas of the former Soviet Union and Western Europe. Less volatile nuclides, including isotopes of strontium and plutonium were deposited primarily within 30 km of the failed reactor.

The dynamics varies with time due to differences in the half-life of the released radioisotopes. In the first few weeks following the Chernobyl accident, the radiation dose to humans was primarily associated with radioiodine. Over succeeding months to years, longer-lived isotopes, particularly ^{137}Cs, with contributions from ^{134}Cs and ^{90}Sr, formed the major part of the dose.

Isotope decay is accompanied by emission of α, β, and γ particles, each with particular toxicological impacts. For example, α particles cannot penetrate the skin but are harmful after they enter body organs via inhalation or ingestion of contaminated food. As β particles cannot penetrate much beyond the skin of humans, β radiation is only harmful after entry into the body; external exposure, however, can, in certain circumstances, cause skin burns. Gamma rays are characterized by significant penetrating power with the capacity to deliver whole-body doses from both external and internal exposures.

The gastrointestinal route is the most significant pathway for the entry of radionuclides in the body. In special circumstances, inhalation and skin contamination may present additional exposures, as for example in the Three Mile Island, Chernobyl and Fukushima accidents. Monitoring and minimizing contamination of food is, therefore, a key strategy in risk management.

An important pathway is the pasture-cow-milk-human route, at least in the immediate aftermath of a contamination event such as the Chernobyl or Fukushima explosions. After this period, other factors particularly relating to the processes involved with the longer-lived particles become relatively more important.

As expected, acute toxicity of absorbed radioactive particles follows proportional dose–response kinetics. Effects range from nausea, vomiting and diarrhoea to internal bleeding, coma and death. At high dosages, radiation can induce instantaneous mortality. Such progressive manifestations of toxicity would be expected to occur in the immediate aftermath of a nuclear attack (as in Hiroshima and Nagasaki) or of accidental explosions (as in Chernobyl and Fukushima).

The accident in Chernobyl resulted in a significant number of cases of acute radiation sickness among plant employees and first responders, but not among evacuees or the general population. Chronic exposure to radioactive particles is unequivocally associated with carcinogenesis at specific sites and processes in the body.

The International Agency for Research in Cancer (IARC) classification of ionizing radiation as a Group 1 carcinogen positions it in a high-risk category. It has recently been estimated that by 2065, the Chernobyl accident will have caused more than 40,000 cases of cancer. Population-based epidemiological evidence amply justifies the IARC classification for ionizing radiation. A number of cancer types are recorded for both *in utero* and direct exposures.

Prenatal exposure to ionizing radiation is associated with elevated risks in offspring, persisting even 30 years later. Pregnancy is considered to be a radiosensitive phase since the fetal thyroid can absorb ^{131}I from the maternal circulation via the placental iodine pump. By late gestation, ^{131}I accumulation in the fetus may exceed that in the mother. Results of a 2019 screening programme in a cohort exposed *in utero* to the Chernobyl fallout indicated increased risks of thyroid cancer and large benign nodules decades later.

Following the Chernobyl accident, the highest incidence of thyroid nodules and cancer occurred among children residing in a particular region (Bryansk) with radiation-contaminated sediment. The increased incidence of thyroid cancer caused by ionizing radiation is due primarily to the emission of radioiodine in the fallout. Iodine is concentrated in the thyroid in response to the demands for thyroxine synthesis by this endocrine gland. The sensitivity of the thyroid gland to radiation-induced cancer is, therefore, entirely predictable. This risk is higher for individuals exposed during infancy and adolescence.

Four types of thyroid neoplasms are recognized:

- differentiated (including papillary) thyroid carcinoma;
- follicular (or vesicular) thyroid carcinoma;
- anaplastic (or undifferentiated) thyroid carcinoma; and
- medullary thyroid carcinoma.

Of these, differentiated thyroid carcinoma is the most common histological type, with an incidence of 80% in people between 30 and 50 years of age.

Extensive screening in Fukushima revealed a high detection rate for thyroid cancer among young individuals. The elevated prevalence of childhood thyroid cancer has been attributed to mass screening and direct comparison with other evidence pertaining, for example, to sporadic/spontaneous cases is made difficult due to differences and limitations in methodology. Particular constraints include the current lack of molecular signatures or genetic biomarkers that would correlate epidemiological evidence with biochemical/physiological mechanisms. Thus, cause-and-effect relationships and dosimetry issues have yet to be established for the increased incidence of thyroid cancer.

Despite the limitations cited above, the search continues for the causes of escalating incidence of differentiated thyroid cancer in children and adolescents after the Chernobyl and Fukushima accidents as well as in the global context (Drozd et al., 2021). Previous observations with atomic bomb survivors indicated that radiation exposure was linked to increased risks for differentiated thyroid cancer, specifically in children. The impetus is now sustained by evidence of genuine worldwide increases in the incidence of differentiated thyroid cancer across all age groups. The Chernobyl data indicate that radiation-induced differentiated thyroid cancer is characterized by a lag phase of 4–5 years prior to diagnosis, with higher incidence in boys and young children and metastases to the lung. Quality of life was markedly restricted due to recurrence of tumours and unavoidable treatment-related side effects.

In contrast, children and adolescents in Fukushima received lower radiation doses to the thyroid and diagnosis of the differentiated cancer type was obtained in population-based screening programmes. The Chernobyl characteristics described above (for example, absence of latency period) were not replicated in Fukushima. However, incidence of tumour stages higher than microcarcinoma and lymph node metastases were noted in the Fukushima cases. Nevertheless, Drozd *et al.* (2021) concluded that there is no definitive proof that radiation played a major role in the pathogenesis of thyroid cancer in Fukushima. Instead, the Chernobyl-Fukushima difference in thyroid cancer incidence was attributed to the potentiating effects of nitrate in drinking water in Belarus compared to Japan. Statistical correlations indicated that the cancer-inducing effects of radiation in Chernobyl were markedly enhanced by nitrates in local drinking water. In contrast, nitrate contamination of water and food was stringently regulated in Japan at the time of the accident and subsequently.

It is well known that differentiated thyroid cancer is a slow progression disorder, albeit with favourable prognosis. However, distant metastases may occur, for example in the lungs and bone. In addition, data published in 2021 indicate the development of rare haemorrhagic brain metastases in differentiated thyroid cancer, affecting the ultrastructure of blood vessels. The underlying mechanism involves a 'rearranged during transfection' proto-oncogene which is affected by exposure to ionizing radiation. It is, therefore, possible that individuals affected by nuclear accidents in Chernobyl, Fukushima and elsewhere may develop clinical variants of differentiated thyroid cancer with haemorrhagic metastases as a latent response to ionizing radiation.

It is now clear that interpretation of epidemiological data is complicated by confounding issues such as disparities in background levels of contaminants such as nitrate. Variations in the application of emerging modalities, screening programmes and over-diagnosis add to these difficulties. In addition, lifestyle factors, particularly incidence of obesity and comorbidity may influence risk assessment.

Problems also arise in the interpretation of data for long-term exposures to radiation. For example, it has recently been observed that solid cancer risks among atomic bomb survivors in Japan remain high more than 60 years after exposure. However, even after using improved dose estimates and adjusting for cigarette smoking, doubts about the shape of the dose–response relationship precludes formulation of reliable and definitive conclusions to inform radiation protection policy makers. Nevertheless, it is consistently stated that the long-term association between exposure to radiation and cancer is well established, with some justification. For example, the occurrence of meningioma among Hiroshima atomic bomb survivors has increased since 1975, with significant correlation between incidence and the radiation dose in the brain.

Recent assessments of the incidence of leukaemia, lymphoma and multiple myeloma among atomic bomb survivors within the Life Span Study cohort, indicated a significant increase in leukaemia risk in Hiroshima and Nagasaki. Ionizing radiation compromises key pathways in haematopoiesis by damaging functional stem cells and the capacity of bone marrow to support regeneration, exacerbated by apoptosis of mature components of the blood. At the molecular level, radiation damages DNA, gene expression and transcription and disrupts signalling networks. The overall clinical outcome of these perturbations is the induction of leukaemia, a critical haematological manifestation, with risks, for example in atomic bomb survivors, persisting throughout the follow-up period up to 55 years after initial exposure to radiation.

Ukrainian children exposed *in utero* to the Chernobyl fallout, showed significant increases for all leukaemia types, including the lymphoblastic form. Similarly, rates of childhood leukaemia were higher in the period after the Chernobyl accident. However, in Sweden, incidence of acute childhood leukaemia did not correlate with degree of radioactive contamination after this accident.

It is well recognized that ionizing radiation increases the risk for the development of breast cancer, particularly when exposure occurs at a young age. In the Life Span Study cohort of atomic bomb survivors covering more than six decades after exposure, results continue to demonstrate a robust dose–response effect for breast cancer incidence in women. However, 2021 data indicate the absence of any increased breast cancer incidence following the Chernobyl accident.

Overall mortality, particularly from non-cancer diseases in impacted populations is another factor of importance in assessing radiation risks. Evidence that radiation affects non-cancer mortality is convincing, with significant increases recorded for cardiovascular, digestive and respiratory causes among atomic bomb survivors. However, emerging surveys suggest that the long-term risks for respiratory disease may be attributed to coincident cancer or other underlying disorders associated with radiation exposure in these subjects.

Other evidence indicated the expression of neurophysiological markers of ionizing radiation. It is relevant here that an updated review on the 20th anniversary of the Chernobyl disaster concluded that mental health effects were the most significant public health issue of this contamination. Among first responders and clean-up personnel with greatest exposure, rates of depression and post-traumatic stress disorder remained high two decades after the explosion. Within the general population, clinical and subclinical depression, anxiety and post-traumatic stress disorder were also recorded.

In conclusion, the incidence of malignancies is consistent with IARC risk assessments of carcinogenesis caused by ionizing radiation. However, it should be noted that the aforementioned events also resulted in the production of other toxic substances, including dioxins and heavy metals. Thus, synergistic interactions may have contributed to carcinogenesis, complicated further by other factors such as iodine status of the soil and nitrates in drinking water.

Nevertheless, mass cancer screening is not advocated for future accidents due to over-diagnosis issues and side effects such as anxiety and over-treatment. However, preparation of contingency measures is essential in view of the worldwide adoption of nuclear power as a carbon-neutral strategy. For example, the commissioning of a new nuclear power plant in Belarus, despite unresolved safety issues, has led the government in Lithuania to offer free iodine tablets to residents living close to the installation. However, local measures should be integrated into a more robust system of environmental impact assessment and international cooperation. Meanwhile, the installation of a new nuclear waste repository in the vicinity of the failed nuclear reactor at Chernobyl underlines toxicological concerns over long-term strategies for disposal of radioactive materials across the globe.

- What are the functions of the thyroid gland?

8.3 Radon

8.3.1. Select (a).

Radon is a radioactive gas formed by the decay of three natural isotopes:

- radon-219, produced from:
  - uranium-235;
- radon-220, due to decay of:

- ○ thorium-232; and
- radon-222, derived from:
  - ○ uranium-238.

With a half-life of 3.8 days, radon emitted from rocks and soil is transported by air and to a limited extent by dissolving in water. Radon gas is stable, fat soluble and does not react with other elements.

Radon gas is artificially concentrated in buildings due to transfer from soil, creating serious health risks for occupants. Radon exposure is increasing within modern residential buildings, attributed to multiple building metrics coupled with lifestyle factors and occupant demographics (Simms *et al.*, 2021).

- Has the term 'decay' (as used above) been explained to you in your course?

8.3.2. A number of features encapsulate continuing concerns over human exposures to radon, including:

- occupational risks;
- epidemiological evidence;
- IARC classification;
- residential exposures;
- lung cancer incidence:
  - ○ histological characterization;
  - ○ lifestyle and occupant demographics as risk factors;
- breast cancer hypothesis;
- potential role in non-malignant conditions:
  - ○ chronic obstructive pulmonary disease (COPD);
  - ○ coronavirus disease (COVID-19) outcomes;
  - ○ neurodegenerative disease;
- relationship in $PM_{2.5}$-associated mortality:
  - ○ the vector concept; and
- diverse mechanisms:
  - ○ need for diagnostics.

Radon exposure has long been associated with lung cancer incidence in mine workers. However, residential exposure is increasingly perceived as an important risk factor in the general population. This is particularly the case in modern buildings in North America and elsewhere, depending on geogenic emissions. Radon exposure is the second leading risk factor for lung cancer after cigarette smoking in the general population.

The IARC classifies radon as a Group 1 carcinogen, unequivocally emphasizing its position as another hazardous environmental gas. This classification is based on evidence obtained with mine workers exposed to high concentrations of radon gas. Low-level radon exposure and lung cancer mortality remains an important issue and continues to attract attention. The relative risk for lung cancer is proportional to cumulative radon exposure. It is estimated that 17% of all lung cancer cases in Alberta (Canada) during 2012 may be attributable to residential radon exposure. There is epidemiological evidence for an increased risk of lung cancer caused by synergism between radon exposure and smoking in mine workers and in the general population.

Health concerns are heightened by expression of different histological lesions following radon exposure. Autopsy analysis of miners exposed to high levels of radon identified an excess of thoracic tumours, classified as lymphosarcoma, but in all probability representing the expression of small cell lung cancer. Subsequently, this form of cancer was also detected in autopsies of uranium miners. This was, therefore, firm evidence of an association between radon exposure and the incidence of proximal lung tumours. Moreover, miners showed all histological types including small cell lung cancer, non-small cell lung cancer, adenocarcinoma and squamous cell carcinoma. Similarly, in one European trial involving the effects of residential radon exposure, all histological types were identified.

Occupant demographics and lifestyle are particular risk factors for radon-induced lung cancer. For example, Simms *et al.* (2021) stated that younger North Americans are exposed to more radon gas associated with occupancy biases within their homes. In addition, modern high radon-containing properties are also more likely to house pregnant women and children. As implied previously, cigarette smoking exacerbates the incidence of this disease. However, emerging evidence confirms that radon is a leading cause of lung malignancy even in never-smokers, with men at higher risk than women (Cheng *et al.*, 2021).

Radon exposure and breast cancer risk has been investigated in the US Nurses' Health Study, with the findings published in 2017. The rationale is based on knowledge that radon gas is lipid soluble and breast tissues receive relatively high doses of radiation. Experimental observations suggest a potential link between radon and incidence of mammary tumours. However, in the Nurses' Health Study, no general association between radon exposure and breast cancer risk was identified, although a suggestive association was proposed for groups with particular oestrogen/progesterone receptor status.

The potential role of residential radon exposure in non-malignant conditions should be considered, particularly for COPD and COVID-19 since inhalation is the primary route of entry, with the lung as the common target organ. Results of a 2020 study highlighted the heterogeneity of available data. However, the authors implied that increases in non-malignant respiratory disease mortality may include evidence of a radon–COPD interaction that might be masked by consideration of other risk factors such as tobacco smoking or inhalation of traffic-related pollutants. Nevertheless, it is hypothesized that chronic exposure to radon may generate a persistent sub-inflammatory pulmonary micro-environment that might promote the onset of COPD. Similar arguments and constraints might apply to proposals of an interaction between radon and COVID-19 outcomes.

In addition, radon exposure may be associated with the incidence or exacerbation of neurodegenerative disorders, as suggested in 2020. The concept is based on the potential of nerve myelin sheath to absorb radon since the gas is fat soluble. Any α-particles emitted may cause irreversible damage to this sheath, with free radicals so generated leading to oxidative damage of myelin.

Other evidence indicates possible interactions between radon and ambient $PM_{2.5}$-related mortality. A time series analysis in 108 USA cities suggested that higher mean city-level radon concentrations increased $PM_{2.5}$-associated cardiovascular and respiratory mortality in the spring and autumn. It is envisaged that radon progenies react with water vapour and ambient pollutants to form highly mobile mixtures which readily coalesce with airborne aerosols. However, although the concept of $PM_{2.5}$ as a vector is plausible, further research is required on the general non-malignant toxicity of radon.

The lack of a coherent hypothesis to explain the underlying mechanisms in radon carcinogenesis adds to ongoing constraints in risk assessment and management. Current proposals include:

- chromosomal degradation;
- gene mutation;
- synthesis of free radicals and cell cycle modification due to production of inflammatory cytokines and proteins;
- oxidative stress induction; and
- epigenetic changes.

For example, chromosomal abnormalities have been observed in miners and non-miners exposed to radon. In addition, gene deletions are known to occur in glutathione *S*-transferases and these abnormalities may increase risk on exposure to radon. Higher incidence of cytogenetic damage in peripheral lymphocytes has been observed in individuals exposed to radon, with the occurrence of polymorphism in certain DNA repair genes, which may represent genetic indicators of chronic radon exposure.

Epigenetic changes in the cell and environmental factors interact in the pathogenesis of lung cancer. In particular, miRNAs have been implicated in the regulation of cellular processes induced by radiation. However, findings in a 2022 review underline the need to elucidate the role of miRNAs in the pathogenesis of radon-induced lung cancer (Kussainova *et al.*, 2022). Establishment of miRNA signatures might assist in identification of high-risk individuals residing in radon-prone environments.

- Explain the abbreviation 'miRNA'.

8.4   Ultraviolet Radiation

8.4.1. Select (a) and (c).

Three types of UV radiation are recognized, based on electrophysical spectra, DNA interactions and human health impacts:

- UVC;
- UVB:
  - photoproducts; and
- UVA:
  - deep penetration into the epidermis.

UVC radiation (at 200–280 nm) presents the highest energy and the lowest penetration capacity in UV radiation. It is absorbed by atmospheric ozone and generally does not reach the earth's surface.

UVB radiation (at 280–320 nm), representing 5% of the UV that reaches the ground, is almost entirely absorbed by the epidermis of the skin. UVB is directly absorbed by DNA inducing molecular rearrangements and forming specific photoproducts. UVB is responsible for diverse forms of damage, representing the most cytotoxic and mutagenic type of solar radiation.

UVA radiation (in the band 320–400 nm), comprises 95% of the solar radiation reaching the earth's surface, penetrating deeply into the human epidermis causing physiological and molecular damage.

In humans, UV radiation is essential for the synthesis of vitamin D in the skin, but excessive exposure is detrimental to health. Perception of solar UV radiation by plants occurs via specialized photoreceptors (Rai *et al.*, 2021).

- Can you name one precursor of vitamin D present in the skin?

8.4.2.   UV radiation is essential for the production of vitamin D which acts separately as a hormone and as a nutrient. This synthesis occurs in the skin, but physiological effects are enhanced after activation in the kidney. Vitamin D promotes absorption of dietary calcium from the gut and contributes to immune functions in the body. Additionally, vitamin D metabolites exert protective functions, particularly instigation of repair responses to offset DNA damage, while promoting apoptosis of malignant cells. Thus, vitamin D aids the cutaneous wound healing process and maintains skin integrity.

Chronic UV overexposure, however, is accompanied by a range of adverse effects. Particular concerns include:

- immunosuppression;
- photoaging;
- inflammation;
- skin cancer:
  - DNA damage;
  - signature mutations;
  - IARC classification;
  - risk factors:
    - previous history;
    - genetics;
    - gender;
    - skin complexion;
    - pre-existing deficits in immunocompetence;
  - squamous cell carcinoma;
  - basal cell carcinoma;
  - cutaneous melanoma, examples:
    - superficial spreading;
    - nodular;
    - lentigo maligna;
  - T-cell lymphoma:
    - mycosis fungoides;
    - Sézary syndrome;
- UV–human papilloma virus synergism; and

- demographics:
  - health alert.

Overexposure to UV radiation is associated with suppression of local immune reactions by damaging and expelling antigen-presenting dendritic cells. In addition, UV radiation attenuates systemic immunity by inhibiting effector and memory T cells. This process enables pre-malignant and tumour cells to escape immune surveillance, thereby facilitating cancer initiation and progression.

Excess UV radiation induces DNA damage and can cause muta-tions of the *p53* gene, which promotes tumour growth and skin can-cer. UV exposure induces gene mutations in key elements of signalling pathways in photoaging ultimately resulting in oxidative damage and inflammation. The IARC classifies UV as a Group 1 carcinogen, empha-sizing the detrimental health risks of overexposure to these photons.

Photo-carcinogenesis conforms with the multi-stage model of can-cer initiation in which UV-induced DNA damage results in mutations that activate oncogenes or silence tumour-suppressing genes. The incidence of skin malignancies increases with decreasing geograph-ical latitude, representing particular risks for fair-skinned individuals in the western hemisphere. Skin cancer is the most rapidly growing malignant disease worldwide, representing health risks particularly for non-melanoma cancer in outdoor workers.

UVB radiation can penetrate into the dermis papillary area and damage DNA in keratinocytes, dendritic cells and melanocytes resid-ing in the skin. UVB carcinogenicity is considerably higher than that of UVA, although this radiation can penetrate the deeper underlying dermal layer of the skin.

Three types of skin cancer are recognized, based on the target cell affected by UV radiation. These include: (i) squamous cell carcinoma; (ii) basal cell carcinoma; and (iii) cutaneous melanoma. The latter occurs less frequently, but is associated with high mortality. Extensive analy-sis of existing data using different mathematical models suggests that the development of squamous cell carcinoma is dependent on total UV exposure, whereas basal cell cancer and particularly cutaneous mela-noma are also dependent upon exposure patterns, with intermittent exposures presenting greater risks.

Squamous cell carcinoma and basal cell carcinoma are often con-sidered together within a 'non-melanoma' classification. These cancers develop gradually in the upper layers of the skin and are more common than cutaneous melanoma. In the UK, for example, approximately 147,000 new cases of non-melanoma skin cancer are diagnosed each year, affecting men more than women as well as the elderly. Basal cell carcinoma is the predominant form of skin malignancy. Non-melanoma skin cancer develops on areas of the body regularly exposed to UV radiation, for example face, hands and shoulders. Although the primary cause of these manifestations is overexposure to UV, other risk factors such as genetics, pale complexion of the skin and pre-existing immunosuppression conditions may predispose to the development of these carcinomas.

Cutaneous melanoma is a more serious form of skin cancer due to the impact of distant metastases. Several types have been identified, based on shape, size and colour of the lesions. Examples include: (i) superficial spreading melanoma; (ii) nodular melanoma; and (iii) lentigo maligna melanoma. Initial signs are the appearance of a new mole or a change in the morphology of an existing growth on the skin. Although the predisposing factors cited above apply in cutaneous melanoma, repeated sunburn episodes increase the risks of contracting this cancer.

The quest for mutational signatures has recently revealed a key role for UV exposure in the development of cutaneous T-cell lymphoma (Jones et al., 2021). Signature 7, implicating UV radiation was uniquely identified in this condition following a meta-analysis of whole exome sequencing data from 400 patients with eight subtypes of T-cell non-Hodgkin's lymphoma. The cutaneous form is a rare type of non-Hodgkin's lymphoma, affecting individuals over 60 years of age with men more likely to develop this condition. Common types of cutaneous T-cell lymphoma include mycosis fungoides and Sézary syndrome. The former is slow progressing, generally only affecting the skin.

In Sézary syndrome, abnormal T cells circulate in the vascular system, impacting the skin as well as other parts of the body, for example the lymph nodes. Sézary syndrome progresses more rapidly than mycosis fungoides.

An emerging feature is the interaction between UV exposure and spread of the human papilloma virus in the development of cutaneous squamous cell carcinoma. It is suggested that the virus is not the primary driver of carcinogenesis but acts as a facilitator in the accumulation of UV-induced mutations.

It is clear that current understanding of the UV–skin cancer relationship is still evolving in terms of both risk factors and underlying pathology. Global incidence of this malignancy has risen markedly over recent years, particularly among young individuals. An alert issued in 2020 that skin cancer mortality in the UK more than doubled in 50 years raises important public health issues concerning general lack of awareness of the dangers associated with excessive UV exposure.

- Have there been alerts in your country about the dangers of excess exposure to UV?

8.5 References

Cheng, E.S., Egger, S., Hughes, S., Weber, M. and Yu, X.Q. (2021) Systematic review and meta-analysis of residential radon and lung cancer in never-smokers. *European Respiratory Review* 30: 200230. https://doi.org/10.1183/16000617.0230-2020

D'Mello, J.P.F. (2020) *Introduction to Environmental Toxicology*. CAB International, Wallingford, UK.

Drozd, V., Saenko, V., Branovan, D.I., Brown, K., Yamashita, S. and Reiners, C. (2021) A search for causes of rising incidence of differentiated thyroid cancer in children and adolescents after Chernobyl and Fukushima: a comparison of the clinical features and their relevance for treatment and prognosis. *International Journal of Environmental Research and Public Health* 18(7), 3444. https://doi.org/10.3390/ijerph18073444

Jones, C.L., Degasperi, A., Grandi, V., Amarante, T.D., Genomics England Research Consortium *et al.* (2021) Spectrum of mutational signatures in T-cell lymphoma reveals a key role for UV radiation in cutaneous T-cell lymphoma. *Scientific Reports* 11, 3962. https://doi.org/10.1038/s41598-021-83352-4

Kussainova, A., Bulgakova, O., Aripova, A., Khalid, Z., Bersimbaev, R. and Izzotti, A. (2022) The role of mitochondrial miRNAs in the development of radon-induced lung cancer. *Biomedicines* 10, 428. https://doi.org/10.3390/biomedicines10020428

Moghadam, Y.K., Dickerson, A.S., Shahedi, F., Bazarfshan, E. and Sarmadi, M. (2021) Association of the global distribution of multiple sclerosis with ultraviolet radiation and air pollution: an ecological study based on GBD data. *Environmental Science and Pollution Research* 28, 17802–17811.

Rai, N., Morales, L.O. and Aphalo, P.J. (2021) Perception of solar UV radiation by plants: photoreceptors and mechanisms. *Plant Physiology* 186, 1382–1396.

Simms, J.A., Pearson, D.P., Cholowsky, N.L., Irvine, J.L., Nielsen, M.E. *et al.* (2021) Younger North Americans are exposed to more radon gas due to occupancy biases within the residential built environment. *Scientific Reports* 11, 6724. https://doi.org/10.1038/s41598-021-86096-3

9 Adaptation in Living Organisms

Questions

9.1 Overview

9.1.1. Adaptation in living organisms exposed to physical and anthropogenic stressors is maximized in: _____, followed in sequence by _____ and _____.

9.1.2. In different species, adaptation to environmental stressors is generally initiated by _____ _____, resulting in upregulation of _____ _____.

9.2 Microbes

9.2.1. Regarding biogenic toxins reviewed in Chapter 2, this volume, microbial activity can confer effects that are:

    (a) beneficial

    (b) detrimental

    (c) potentiating

    (d) none of the above

9.2.2. Are microbes endowed with the capacity to metabolize persistent organic pollutants (POPs) and other environmental contaminants?

9.2.3. Explain why microbial adaptation is deleterious in clinical settings and may also impact on pre-existing human health disorders that are exacerbated by toxic air pollutants.

© J.P.F. D'Mello 2022. *Key Questions in Environmental Toxicology* (J.P.F. D'Mello)
DOI: 10.1079/9781789248548.0009

9.3 Higher Plants

9.3.1. The pH of acid rain is:

(a) about 14

(b) between 7 and 11

(c) between 12 and 14

(d) between 5 and 6

9.3.2. Explain how plants respond and adapt to acid rain and other environmental stressors.

9.4 Animals

9.4.1. 'Killifish'.

Using the article by Whitehead *et al.* (2017), prepare a case study on evidence of adaptation to pollutants and contaminants by animals.

Answers

9.1 Overview

9.1.1. Insert, respectively: microbes; plants; animals.

All living organisms are endowed with mechanisms that facilitate adaptation to changes in the type and levels of stressors present in the environment. These processes permit microbes, plants and animals to resist or even eliminate adverse effects, with varying degrees of success. Adaptation is embodied in the concept of 'natural selection' as used in evolutionary biology. An additional concept based on 'evolutionary rescue' has now emerged to imply adaptation in living organisms exposed to complex mixtures of pollutants.

A number of factors interact in the expression of adaptive responses to physical and anthropogenic stressors including, but not limited to:

- the nature of the stressor;
- membrane properties;
- nucleotide diversity;
- genetic variation;
- genome size;
- horizontal gene transfer;

- generation time; and
- population size.

Optimistic indications regularly appear in scientific journals about the potential for bioremediation based on naturally occurring microbes in contaminated sites. The term 'microbial extremophiles' has been coined for such organisms that are capable of adapting to hostile environments and degrading a wide range of organic pollutants.

- Explain the significance of 'horizontal gene transfer'.

9.1.2. Insert: gene expression; metabolic (or 'detoxification') pathways.

The initial event invariably involves gene expression, followed by stimulation of enzymes responsible for detoxification or sequestration of the xenobiotic compound. For example, salinity generates reactive oxygen species (ROS) which precipitates oxidative stress in plants. Gene expression follows, resulting in the upregulation of pathways leading to the synthesis of signalling compounds such as salicylic acid which stimulate the production of protective osmolytes, particularly amino acids.

A similar sequence of gene expression and upregulation of metabolic pathways occurs in microbial adaptation to chemicals. Relevant examples are the development of antibiotic resistance in bacteria and fungicide resistance in pathogenic fungi.

- Can you define 'signalling compounds'?

9.2   Microbes

9.2.1. Select (a) and (b).

Microbial activity can confer both beneficial and detrimental effects following the metabolism and final disposition of biogenic toxins. These effects are exemplified in mechanisms of microbial adaptation in the rumen of cattle and sheep. Key issues relating to toxic amino acids include:

- anaerobic conditions in rumen;
- fermentation by microbes;
- production of reactive metabolites;
- forage composition; and
- contrasting physiological manifestations.

The toxicity of the non-protein amino acid, S-methylcysteine sulfoxide, present in forage brassicas, is mediated only after metabolism by rumen microbes to a reactive metabolite, dimethyl disulfide. This product precipitates a severe form of haemolytic anaemia within 1–3 weeks in ruminants fed mainly or exclusively on forage brassicas. Thus, this toxicity only occurs via the intervention of specialist microbes adapted to operating under anaerobic conditions in the rumen.

In contrast, the toxicity of mimosine in *Leucaena leucocephala* forage is determined by variations in rumen microbial ecology in cattle, sheep and goats. The adverse effects are critically dependent upon the rate and extent of bacterial degradation of mimosine to its metabolite, 3-hydroxy-4(1*H*)-pyridone and a related isomer. Initial results implied that ruminants in Australia, the USA and Kenya lacked the requisite bacteria to complete the degradation of these metabolites and, consequently succumb to goitrogenic effects if relatively high intakes of *Leucaena* are maintained over protracted periods of time.

In other regions where *Leucaena* was indigenous (Central America) or naturalized (Hawaii and Indonesia), ruminants there possessed the full complement of bacteria that are required for degradation of the mimosine metabolites, which accounted for the absence of *Leucaena* toxicity in these countries. However, it now appears that the detoxifying bacteria are ubiquitous in distribution, rather than restricted to geographical locations (Mcsweeney *et al.*, 2019).

• Define 'anaerobic'.

9.2.2. There is increasing evidence that microbes are capable of metabolizing POPs and other pollutants in temperature-dependent reactions. Microbes stimulate biodegradation of organic compounds via several mechanisms, including:

• constitutional alterations in microbial consortia (it is instructive to note that in a 2017 study on Norwegian seawater and sediments, degradation of polychlorinated biphenyls (PCBs) was accomplished by genera not previously reported to possess this capacity);
• abundances of particular genera;
• mutations;
• horizontal gene redistribution:
  ○ genome size;
• epigenetic modifications; and
• recombinant occurrences.

Although there is reliable evidence indicating that microbes are endowed with biochemical mechanisms to metabolize POPs at contaminated sites, it is salutary to note important constraints about this

157

capacity. For example, contaminated sediments from a wastewater lagoon in Virginia (USA) contained microbial communities dominated by *Proteobacteria*, *Firmicutes* and *Clostridia* with the potential to dechlorinate PCBs. Data indicated the presence of PCB reductive dehalogenase genes to support such a mechanism. Other evidence points to the microbial degradation of dioxins and furans at a site with multiple contaminants in Oregon (USA). Despite these favourable results, it is important to express a cautionary note in view of earlier data indicating that dehalogenation of highly chlorinated dioxins by anaerobic microbial consortia in river sediments may result in products that are potentially more toxic than the original parent compounds.

Genome size in microbial communities may determine degradation efficacy following exposure to POPs. Dong *et al.* (2021) examined this hypothesis in responses of bacterial taxa to organic pollutants, with the observation that species with larger genome size are more adaptable to stress imposed by polycyclic aromatic hydrocarbons (PAHs). It is argued that horizontal transfer of degrading genes across microbial communities is critical for adaptation to pollutants in the environment.

There are clear indications of biodegradation and greater expansion of microbial diversity in marsh sediments impacted by the *Deepwater Horizon* accident in the Gulf of Mexico. It is generally concluded that microbial activity has contributed to the ongoing recovery in the impacted ecosystem. The role of microbial extremophiles in petroleum hydrocarbon degradation is now emerging as a potential bioremediation process.

Studies on the microbial degradation of plastic debris in waterways and oceans is ongoing. Initial evidence indicates low distribution of microbes with polyethylene terephthalate hydrolases in marine and terrestrial ecosystems. It is important, however, that transgenics are not used to enhance capacity in the natural environment, due to disquiet among environmentalists and the public in general about the ethics and safety of these methodologies.

The evolution of arsenic resistance has been implied by observations that specialized bacterial extremophiles are able to tolerate extremely high levels of arsenic in an alkaline lagoon nestling within the caldera of an active volcano in Argentina. Concentrations of arsenic in the water are 20,000 times the level regarded as safe for drinking water.

- What is 'dehalogenation'?

9.2.3. Microbial adaptation is deleterious in clinical settings for two principal reasons, namely: (i) development of multidrug resistance; and (ii) exacerbation of pulmonary disorders, particularly in patients compromised by exposure to harmful air pollutants. Salient issues in these forms of microbial adaptation are summarized below:

- multidrug resistance:
  - genetic factors;
  - epigenetic effects;
  - accumulation of resistant genes; and
- microbial exacerbation of pulmonary disorders linked to toxic air pollutants, including:
  - role of opportunistic microbes:
    - transition from commensal to invasive mode;
    - *Escherichia coli*;
    - *Aspergillus fumigatus*;
    - *Candida albicans*;
  - cystic fibrosis, with particular:
    - disease characteristics;
    - gut microbiome features;
    - aspergillosis;
  - chronic obstructive pulmonary disease (COPD) and asthma, impacted by:
    - air pollution;
    - allergic bronchopulmonary aspergillosis;
  - coronavirus disease (COVID-19) infection:
    - aspergillosis;
    - candidiasis (including evolution of invasive forms of *Candida albicans*).

An issue of particular concern, to environmentalists and clinicians alike, is the acquisition of antibiotic resistance by bacteria (see also Chapter 7, this volume). Two examples of pathogens relevant in clinical settings include penicillin-resistant *Streptococcus pneumoniae* and methicillin-resistant *Staphylococcus aureus* (MRSA).

Bacteria can acquire antibiotic resistance directly through genomic mutations. Alternatively, antimicrobial resistance may occur via horizontal gene transfer, usually involving mobile genetic elements (Johansson *et al.*, 2021). These elements enable the spread of antibiotic resistance throughout microbial species and communities. Resistance in bacteria evolves not only as a result of an accumulation of mobile genetic elements, but also in response to epigenetic modifications, enabling some pathogens to switch to an antibiotic-tolerant dormant mode. In *Escherichia coli*, for example, extremely high levels of multidrug tolerance can be established by single point mutations in one of several genes, whereas reversion to low persistence mode in the absence of antibiotic can be relatively slow.

Microbial adaptation also impacts on pre-existing human health disorders that are exacerbated by toxic air pollutants. For example, opportunistic bacteria and fungi are able to colonize and proliferate in the lungs, thereby interacting with air pollutants to aggravate manifestations and other characteristics in these disorders.

A cogent example of adaptation in bacteria concerns commensal species of *E. coli* in the intestinal tract of young children with cystic fibrosis. This disorder is characterized by pulmonary and intestinal inflammation as well as infection, resulting in significant morbidity and premature death.

Although cystic fibrosis is essentially a genetic disorder, there are environmental dimensions in manifestations and final outcomes, particularly in relation to exposures to ambient air pollutants. Matamouros *et al.* (2018) maintains that *E. coli* generally represents less than 1% of the human gut microbiome. However, in cystic fibrosis patients, in excess of 50% relative abundance is common, corresponding with Intestinal inflammation and faecal fat malabsorption. Matamouros *et al.* (2018) demonstrated that *E. coli* from young children with cystic fibrosis can adapt to utilize glycerol, a major component of faecal fat. Patients with cystic fibrosis may also develop allergic bronchopulmonary aspergillosis in response to infection caused by the fungus *Aspergillus fumigatus* (Chapter 2, this volume).

It is widely acknowledged that both bacteria and viruses are adapted to proliferate within the respiratory tract microbiome, aggravating underlying conditions such as COPD and asthma. Emerging evidence implicates ambient particulate matter and oxides of nitrogen in the increased incidence of COPD exacerbations, accompanied by infective aetiology.

A number of risk factors are linked to asthma exacerbations, particularly in children. Genetics and exposure to allergens, traffic-related pollutants and tobacco smoke are major determinants in the severity of manifestations displayed by these children. However, it is consistently claimed that viral infections are critical in aggravating symptoms and clinical outcomes in young asthmatics.

The COVID-19 pandemic has revealed a further example of adaptation in host–microbial interactions, compounded by toxic air pollutants. There are persistent claims that residents in polluted city centres are at greater risk of severe manifestations and mortality in COVID-19 infections compared to those living in suburban areas.

Thus, it is clear that mechanisms of adaptation exist in microbes to cope with different challenges associated with antibiotic overuse. In combination with noxious ambient pollutants, transition of bacteria and fungi from commensal to invasive forms accelerates onset or exacerbates effects in cystic fibrosis, COPD, asthma and COVID-19 syndromes.

- What are 'commensal' microbes?

9.3 Higher Plants

9.3.1. Select (d).

Acid rain remains the pre-eminent example of pollution caused by anthropogenic activity, with widespread impacts on different ecosystems. It is caused by the deposition of acidic components of rain, snow, hail and fog with a pH less than 5.6. In general, pH values in the range 4.4–2.3 have been recorded, but a value of 1.4 has been reported in Italy.

Occurrence of acid rain is a particular issue in China, attributed to industrial pollution, where pH values ranging from 4.0 to 8.6 were recorded in the source area of the Yangtze River during the 2010–2015 period. The transboundary nature of acid rain pollution was emphasized, with sources attributed to India. At the regional level, local topography and prevailing air currents will define the area that is impacted by acid rain.

Acid rain is linked to global emissions of sulfur dioxide and nitrogen oxides into the atmosphere. The combustion of fossil fuels, particularly coal, accounts for a significant proportion of this pollution. It has recently been reported that the composition of acid rain is gradually changing from sulfuric to nitric acid. As might be expected, acid precipitation is determined by complex dynamics involving other pollutants. For example, it is argued that ammonia emission might mitigate haze pollution and nitrogen deposition, but exacerbate acid rain impacts.

- What is 'haze' pollution?

9.3.2. The emergence of herbicide resistance marks a defining example of adaptation in higher plants, in terms of speed and extent of evolution (Baek *et al.*, 2021). The effects of other stressors, including acid rain, drought, salinity, hypoxia/anoxia and toxic metals should be measured against this perspective.

Plants respond and adapt to abiotic stressors in different ways summarized as follows:

- responses to:
  - acid rain:
    - morphological impairments;
    - decreased chlorophyll synthesis;
    - rhizosphere impairments;
    - decreased plasma membrane function;
    - increased mobility of heavy metals;
  - drought:
    - loss of productivity;
    - reduced survival;
  - salinity:
    - osmotic stress;
    - ionic toxicity;
    - reduced uptake of essential minerals;
    - functional disruption in root cells;
    - water loss;
  - hypoxia/anoxia:
    - energy crisis;
    - reduced productivity and survival;
  - toxic metals, of particular relevance in serpentine soils:
    - decreased rates of nutrient uptake;
    - reduced tissue turnover;
    - impaired leaf growth;
    - larger flower heads; and
- mechanisms of adaptation:
  - signal perception and transduction;
  - expression of stress-responsive genes:
    - effects of transcription factors;
  - upregulation of specific enzymes;
  - synthesis of key phytohormones and other signalling macromolecules;
  - deployment of regulators:
    - aquaporins;
    - osmolytes (e.g. raffinose and proline);

- morphological alterations;
- complexation and sequestration:
  - role of amino acids;
  - phytochelatins.

Acid rain destroys the waxy cuticle of the leaf surface, damages the epidermal structure, with the constituent acids diffusing into the cell or via the stomata. Leaf integrity is critical for maintaining water balance in the plant and for gaseous exchange. It is also important for protection against pathogen invasion and entry of other pollutants. Visible leaf injury indicators, including appearance of necrotic spots, defoliation and discoloration are not evident until pH values fall to below 4.0, with the size of necrotic spots increasing with decreasing pH.

Leaf chlorophyll content directly reflects foliage damage and consequent loss of plant productivity caused by acid rain. Chlorophyll synthesis is decreased by acid rain due to foliar leaching of nutrient elements, particularly magnesium, one of the major components of the pigment. Degradation of chlorophyll is also increased by acid rain, leading to the formation of pheophytin.

Acid rain affects the rhizosphere by changing the pH value of the soil and altering the availability of nutrients. Root morphology, as expressed as root length, surface area, volume and number of root tips are all reduced by strongly acid rain (e.g. pH 2.5), arising partly from decreased macro-elements and increased trace element status in roots.

There is increasing evidence that acid rain increases levels of ROS which precipitates peroxidation and loss of membrane integrity, thereby disrupting cellular ionic status and functional metabolism. Oxidative stress is a critical and common manifestation of exposure to acid rain as well as other abiotic factors, although plants possess adaptive defence mechanisms to counteract this effect.

Acid rain also increases the mobility of heavy metals in soil and their uptake by plants. In addition, ambient ozone may interact with acid rain to reduce net photosynthetic rates. The effects depend on species of plants, environmental factors and concentrations of pollutants.

The overall effects of drought, salinity, hypoxia/anoxia and toxic metals are reflected in reduced growth and survival of plants, mediated via a variety of pathways. For example, salinity is associated with osmotic stress, ionic toxicity, impaired uptake of essential minerals and water deficit, resulting in reduced productivity, particularly in crop plants.

The effects of toxic metals are ecologically relevant due not only to industrial contamination but also to the natural and worldwide occurrence of serpentine soils with relatively high concentrations of nickel, cobalt and chromium. Despite these characteristics, spontaneous evolution and speciation of serpentinophytes are frequently observed in diverse locations. Species include strict metallophytes as well as facultative populations with high genetic variation and potential for adaptation to these extreme conditions.

Adaptation to drought is the result of multifactorial processes in plants, involving proteins, transcription factors and other mediators. For example, aquaporins act as membrane channels, exerting important roles in cellular water and osmotic homeostasis in plants and adaptation to water-deficit conditions. Aquaporins are also linked to acquisition of carbon, nitrogen and micronutrients in plants.

Twenty-one aquaporin genes have been identified in *Acacia auriculiformis* by Zhang *et al.* (2021). Expression profiles of these genes occur in response to drought and salinity. The role in adaptation was underlined in *Arabidopsis* plants overexpressing one of the *Acacia* aquaporin genes, resulting in reduced water loss. In *Canavalia rosea*, expression of aquaporin genes has also been implicated in adaptation to stress imposed by drought and saline–alkaline conditions.

In other plants, dehydration-induced transcription factors are associated with enhanced drought and salinity tolerance. It has been proposed that these transcription factors, localized in the nucleus, can induce tolerance to drought and osmotic stresses by promoting root growth and water retention as well as enhancing the antioxidant enzyme system, ROS homeostasis and facilitating membrane-lipid integrity.

Adaptation to salinity involves the expression of ROS-scavenging enzymes including superoxide dismutase, peroxidase and catalase. Non-enzymatic mechanisms based on the production of antioxidants such as ascorbic acid and phenolic compounds are also implicated.

Other intermediates associated with adaptation to salinity include raffinose and proline, classified in this context as 'osmolytes'. Sugarbeet is recognized as a salt-tolerant crop capable of growing in degraded saline soils. Emerging evidence suggests that accumulation of raffinose in leaves confers salinity tolerance in these plants via action as an osmolyte or scavenger of ROS and other toxic radicals (see review by Chourasia *et al.*, 2022).

Proline accumulation is widely regarded as a mechanism for adaptation of plants to abiotic stressors. Its efficacy is associated with specific properties, including osmoprotection, regulation of cellular homeostasis, signal transduction and gene expression. In addition, proline acts by providing structural stability to proteins, including enzymes, membranes and key organelles such as chloroplasts and mitochondria.

Waterlogging is a serious issue worldwide, resulting in temporary or prolonged oxygen stress in submerged plants. The terms 'hypoxia' and 'anoxia' are used in this context, with particular reference to oxygen status in the rhizosphere. In response to such constraints, some plants adapt via morphological mechanisms to increase supply of oxygen. For example, in several species, development of additional aerenchyma and adventitious roots has been observed, aimed at improving oxygen status.

Mechanisms exist in plants to reduce the impact of deleterious metal ions, through the mediation of specific amino acids and peptides. Proline represents a prominent exponent of amino acids with metal-complexing attributes, based on biophysical properties. The role of histidine in nickel hyperaccumulation in certain plants of the genus *Alyssum* is also well established. Coordination of nickel by histidine has been demonstrated and it is suggested that this amino acid serves as a ligand for the element, thereby promoting hyperaccumulation. In populations of *Silene paradoxa*, adaptation to serpentine soils containing high levels of nickel is attributed to a metal excluder strategy.

Relatively efficient and complex mechanisms exist in plants for chelation, transport and vacuolar sequestration of toxic metals within plant cells through the action of phytochelatins. The phytochelatins are a group of cysteine-rich peptides synthesized exclusively from reduced glutathione. It is instructive to consider the model recently proposed for phytochelatin-mediated arsenic tolerance in plants. It is proposed that soil arsenic is absorbed and transported by a high-affinity phosphate transporter into root cells. Arsenic reductase reduces arsenic (V) to the (III) form which is conjugated by phytochelatin and transported by ABCC1 and ABCC2 transporters to vacuoles where it is sequestered, thereby inducing arsenic tolerance in the plant.

Thus, a diverse range of adaptation mechanisms exist in plants to circumvent the adverse effects of abiotic stressors. This situation reflects the well-defined molecular evidence underlying the evolution of herbicide resistance in weeds.

- Can you explain the difference between hypoxia and anoxia?

9.4 Animals

9.4.1. Case study: 'Killifish'

It is generally assumed that adaptation to environmental contaminants might be a protracted and restricted process in the animal kingdom. However, it is pertinent to reflect on the pace of development of insecticide resistance in arthropod pests.

If insects can adapt to toxic chemicals, then it is relevant to enquire whether higher animals also possess mechanisms to process environmental contaminants and reduce deleterious effects. Key issues under consideration include:

- the killifish evidence:
  - ecology;
  - contaminants;
  - elements of 'evolutionary rescue':
    - genetic variation;
    - population size;
    - generation turnover;
- turtle evidence:
  - 'evolutionary trap';
- prospects for endangered marine predators;
- human adaptation:
  - arsenic;
  - comparisons with archaic hominins; and
- a reality check.

Killifish (*Fundulus heteroclitus*) are able to survive polluted waters in a variety of settings. For example, the aquatic ecosystem in the New Bedford Harbor in Massachusetts is renowned for its content of toxic compounds, particularly PCBs deposited between 1940 and 1970 through industrial activity. Despite this level of contamination, the Harbor is home to a thriving population of killifish. Similarly, Gulf killifish populations in Houston (USA) have survived polluted waters to make a remarkable recovery.

The concept of 'evolutionary rescue' is gradually emerging in the discussion over adaptation in animals to environmental contaminants. In the case of killifish, a number of factors may contribute to survival and expansion of the population in the two locations cited above. According to Whitehead *et al.* (2017), the primary elements driving adaptation include genetic diversity, population size and generation turnover rates. It is suggested that genetic mutations involving inter-species mating with Atlantic killifish have provided vital nucleotide

Continued

9.4.1. Continued.

diversity to promote adaptation to pollutants. A large population size is also essential to provide the genomic variation implicit in this process. Relatively short generation times also play an important role in adaptation. Thus, killifish live for just 3 years, providing opportunities for adaptation over a significant number of generations.

Whitehead *et al.* (2017) strike an optimistic note in suggesting that evolutionary impact of anthropogenic stressors occurs widely to facilitate adaptation in animals. Furthermore, the notion that 'evolution is the solution to pollution', as incorporated in the title of the said paper, may also convey a positive message.

However, for a considerable number of species impacted by common pollutants, the imposition of an 'evolutionary trap' is more appropriate. This term has specifically been applied to the plight of young sea turtles in the Pacific and Indian Ocean coasts of Australia noted for high levels of plastic pollution. According to a 2021 report, these reptiles travel on ocean currents as part of evolved behaviour but ingest plastic polymers instead of food, thus leading to significant declines in populations.

In the case of apex predators, prospects of adaptation to industrial pollutants are severely restricted. Declines in populations of polar bears and killer whales have been predicted due to chronic exposures to PCBs, chlorinated pesticides, perfluoroalkyl residues, dioxins and toxic metals. In combination with climate change and habitat loss, these contaminants create existential risks for a diverse range of marine predators.

Remarkably for a toxic metal, there is evidence of adaptation to arsenic in human populations. Communities in the northern Argentinian Andes are routinely exposed to arsenic via drinking water. Emerging evidence indicates that individuals there are endowed with unique mechanisms of metabolism that permit tolerance to this toxic element (for details of the toxicology of arsenic in humans see Chapter 6, this volume).

The evolution of hominin detoxification as reviewed by Aarts *et al.* (2021) is worthy of consideration. It was hypothesized that regular exposure to harmful constituents in smoke might instigate adaptation in genes involved with tolerance towards these particles. Further analysis suggested that archaic hominins had predominantly protective gene variants, as do extant chimpanzees and gorillas. However, the number of less protective, 'high risk' alleles has increased in modern humans.

A reality check is appropriate here. For beneficial insects, apex predators and vulnerable individuals exposed to toxic pollutants, evolutionary rescue is an unlikely outcome.

- What are 'hominins'?

9.5 References

Aarts, J.M.J.G., Alink, G.M., Franssen, H.J. and Roebroeks, W. (2021) Evolution of hominin detoxification: Neanderthal and modern hominin Ah receptor respond similarly to TCDD. *Molecular Biology and Evolution* 38, 1292–1305.

Baek, Y., Bobadilla, L.K., Montgomery, J.S., Murphy, B.P. and Tranel, P.J. (2021) Evolution of glyphosate-resistant weeds. *Reviews of Environmental Contamination and Toxicology.* https://doi.org/10.1007/398_2020_55

Chourasia, K.N., More, S.J., Kumar, A., Kumar, D., Singh, B. *et al.* (2022) Salinity responses and tolerance mechanisms in underground vegetable crops: an integrative review. *Planta* 255: 68. https://doi.org/10.1007/s00425-022-03845-y

Dong, Y., Wu, S., Fan, H., Li, X., Li, Y. *et al.* (2021) Ecological selection of bacterial taxa with larger genome sizes in response to polycyclic aromatic hydrocarbon stress. *Journal of Environmental Sciences* 112, 82–93.

Johansson, M.H.K., Bortolaia, V., Tansirichaya, S. and Petersen, T.N. (2021) Detection of mobile genetic elements associated with antibiotic resistance in *Salmonella enterica* using a newly developed web tool: mobile element finder. *Journal of Antimicrobial Chemotherapy* 76, 101–109.

Matamouros, S., Hayden, H.S., Hager, K.R., Brittnacher, M.J., Lachance, K. *et al.* (2018) Adaptation of commensal proliferating *Escherichia coli* to the intestinal tract of young children with cystic fibrosis. *Proceedings of the National Academy of Sciences of the United States of America* 115, 1605–1610.

Mcsweeney, C.S., Padmanabha, J., Halliday, M.J., Hubbard, B., Dierens, L. *et al.* (2019) Detection of *Synergistes jonesii* and genetic variants in ruminants from different geographical locations. *Tropical Grasslands – Forrajes Tropicales* 7(2), 154–163. https://doi.org/10.17138/tgft(7)154-163

Whitehead, A., Clark, B.W., Reid, N.M., Hahn, M.E. and Nacci, D. (2017) When evolution is the solution to pollution: key principles and lessons from rapid repeated adaptation of killifish (*Fundulus heteroclitus*) populations. *Evolutionary Applications* 10(8), 762–783.

Zhang, G., Yu, Z., Teixeira da Silva, J.A. and Wen, D. (2021) Identification of aquaporin members in *Acacia auriculiformis* and functional characterization of *AaPIP1-2* involved in drought stress. *Environmental and Experimental Botany* 185: 104425. https://doi.org/10.1016/j.envxpbot.2021.104425

10 Discussion

Questions

10.1 Overview

10.1.1. The three environmental emergencies attributed to pollution include: _____ _____, _____ _____ _____ and _____.

10.1.2. The three principal human health disorders exacerbated by environmental stressors include: _____ _____, _____ _____ and _____.

10.1.3. Complete the following titles of monitoring organizations cited in this volume or based on your course lectures and supplementary reading:

    (a) Environmental Protection _____ (EPA; USA).

    (b) Food and _____ _____ (FDA; USA).

    (c) _____ Health Organization (WHO; United Nations).

    (d) International _____ for Research in _____ (IARC; WHO).

10.1.4. As requested in the Preface, this volume, submit here your environmental diary for the past 9–12 months with information recorded in chronological order, including toxicological implications of recorded incidents or announcements. It is recommended that you present the information in a suitably structured table.

10.2 Distribution of Contaminants and Pollutants

10.2.1. Write an essay on ambient air pollution.

10.2.2. What are the important issues in soil pollution?

10.2.3. The claim that the Mediterranean Sea is 'dying' appeared in 2019. Discuss the global issue of pollution in marine and freshwater ecosystems in the light of that announcement.

10.2.4. Evaluate the role of biomonitoring procedures in assessing the quality and safety of surface waters.

10.2.5. Discuss how human health may be compromised by the occurrence of contaminants and pollutants in foods.

10.3 Human Health Emergency

10.3.1. Discuss the emerging public health issues associated with ambient air pollution in a city of your choice, including reference to current concerns over residential exposure in other major metropolitan locations.

10.3.2. Using specific examples, evaluate the basis of pathogen–pollutant interactions in determining outcomes in patients with underlying pulmonary disorders.

10.3.3. Discuss the role of genetic factors in human health responses to environmental contaminants.

10.3.4. Explain how morbidity in obese individuals may be exacerbated by exposure to environmental pollutants.

10.3.5. Compare and contrast the human health impacts of named environmental carcinogens.

10.3.6. Discuss how tobacco smoking and environmental stressors may interact to affect incidence of lung cancer.

10.3.7. A 2021 news item indicated that expectant women exposed to air pollution 'were more likely to have children with asthma'. Use this statement and epidemiological evidence to prepare a case study on 'pollution and pregnancy outcomes in women'.

10.3.8. Discuss the neurological impairments associated with exposure to environmental stressors.

10.3.9. Explain the relevance of 'oxidative stress' and 'inflammation' in human morbidity linked to pollution.

10.3.10. Discuss the extent to which pollutants may interact with cellular receptors and organelles to determine human health outcomes.

10.4 Ecological Emergency: Biodiversity in Peril

10.4.1. Discuss the prediction that an 'insect apocalypse' is only a matter of time. You should focus on the likely impacts of named pollutants.

10.4.2. Using specific examples, explain how pollutants can affect populations of amphibians.

10.4.3. Write an essay on the impact of pollutants on populations of avian species.

10.4.4. There is considerable disquiet that populations of predators may be at severe risk due to pollution in diverse habitats and ecosystems. Discuss the evidence for this concern.

10.5 Risk Assessment and Regulation

10.5.1. By means of a table summarize the toxicological risks associated with pollution at various levels of human organization and activity. Add a short commentary in the table to explain salient points. Use your lecture notes and supplementary reading to complete this table.

10.5.2. Define 'epigenetics' and discuss its relevance as a component of risk assessment systems in environmental toxicology.

10.5.3. Outline the system currently in place to monitor and control emissions and distribution of toxic pollutants and discuss its effectiveness in minimizing risks for human health and threatened wildlife species.

10.5.4. Using specific examples, critically evaluate the risk assessment methodologies adopted by major internationally recognized agencies in the regulation of toxic pollutants.

10.5.5. Outline and critically evaluate the toxicological criteria used by regulatory authorities to establish the risk profile of a named pesticide prior to final approval.

10.6 Policy

10.6.1. 'Levelling up' is a political slogan and promise that is now lauded and derided in equal measure in the UK and elsewhere. Discuss the extent and significance of environmental inequalities that require attention to ensure delivery on this pledge.

10.6.2. It has been suggested that greater reliance on plant-based foods will contribute significantly to reducing the detrimental impact of anthropogenic activity on the environment. Discuss how implementation of this policy would result in unintended consequences for human health and biodiversity.

10.7 Compliance

10.7.1. Outline the protocol used by US EPA to ensure compliance with pollution regulations and directives. Use specific examples to explain the major features of this system.

10.7.2. Discuss why vigilance should be an important strategy in ensuring compliance with environmental regulations; you are advised to use examples from your diary to support your arguments.

10.8 Conclusions

10.8.1. Summarize your main conclusions regarding the toxicological implications of anthropogenic activity.

10.8.2. In the light of evidence that you have read in this volume and gathered from your course curriculum in environmental toxicology, what are your recommendations to protect human health and wildlife species on the brink of extinction?

Answers

10.1 Overview

10.1.1. Insert: climate change; increased human morbidity; ecotoxicity. Other words may be equally suitable, for example 'diseases' and 'biodiversity loss', with appropriate changes to the question. One of the key issues to convey is the path of carbon in the three environmental emergencies, emphasizing the imperative of extending global objectives in the pursuit of carbon neutrality. Cross-linking in these networks exemplifies the impact of climatic factors in the generation of contaminants and pollutants harmful to human health and wildlife species.

• Can you explain the term 'carbon neutrality'?

10.1.2. Insert: lung diseases; cardiovascular disease; cancer.

Again, greater flexibility should be allowed in this answer, to include 'pulmonary' and 'heart' in place of 'lung' and 'cardiovascular'. The principal lung diseases exacerbated by pollution include asthma, chronic obstructive pulmonary disease (COPD) and cystic fibrosis. Similarly, cardiovascular disorders relevant in this context include hypertension and cardiac arrest. Environmental stressors are associated with different forms of cancer, affecting the lungs, liver and the thyroid gland, depending on the nature of the stressor.

- What is another term for 'cardiac arrest'?

10.1.3. Insert:

(a) Agency.

The Environmental Protection Agency (EPA) in the USA aims to 'protect human health and the environment'. This objective includes ensuring that Americans have clean air, water and land. The EPA coordinates national efforts to reduce environmental impacts based on the available scientific data. It also implements federal laws to protect human health equitably and as effectively as intended by the US Congress.

In discharging these duties, the EPA aims to ensure that stakeholders and citizens are provided with access to accurate information and advisories. It also works to ensure that contaminated lands and toxic sites are reclaimed after decontamination.

(b) Drug Administration.

The Food and Drug Administration (FDA) in the USA imposes regulations to ensure quality and safety of bottled water, dietary supplements, food additives, potential allergens, prescription and non-prescription drugs, cosmetics and veterinary products.

(c) World.

The World Health Organization (WHO) is responsible for directing and coordinating international health programmes, operating under the auspices of the United Nations. In addition to overseeing international health systems, the WHO conducts surveillance on non-communicable and communicable diseases and coordinates responses to outbreaks and epidemics.

(d) Agency; Cancer.

The International Agency for Research in Cancer (IARC) is a subsidiary of the WHO. It coordinates and undertakes epidemiological and laboratory-based research to establish causes, risks and mechanisms of carcinogenesis in human populations worldwide.

- Why is it important to obtain mechanistic data?

10.1.4. The author's annotated diary is presented in Table 10.1.

Table 10.1. The author's environmental diary relating to reported incidents or announcements, with comments on and cross-referencing to toxicological risks. Reports in the media are generally based on evidence published in scientific papers.[a]

Month and year	Incident or announcement	Comments/toxicological implications
June 2020	'Levelling up'	A political pledge promulgated in the UK, but social and ethnic inequalities in exposure to vehicular and industrial air pollution remain a formidable issue worldwide
	Extensive dust plume drifts into the USA	This event occurs every year and routinely creates respiratory illness in affected populations; can also trigger toxic algal blooms
	Microfibres in textiles polluting oceans	Polyester, acrylic and nylon fibres released during laundering of clothes
	Mediterranean plastic pollution hotspots revealed	Major contributors: Egypt, Turkey and Italy
	Some imported plastic waste burned on roadsides in Turkey	Emissions of styrene gas, dioxins and particulates are risk factors
	'Arctic on fire' (repeated in September 2020)	Evidence depicted in satellite images; risks due to particulates and PAHs for residents 20 km from wildfire
	Russian mining conglomerate admission on wastewater spill in the polar Arctic	Potential for ecotoxicological risks
July 2020	Major UK oil company fined £7000 for crude oil spill into the North Sea	A paltry fine for adding to contamination in marine ecosystems
	'Mine more nickel'; urges chief executive of a major electric vehicle manufacturer	This exhortation should assist in the move away from the combustion of fossil fuels, in favour of increased use of electric vehicles; however, Ni mining and disposal of used batteries creates toxicological problems (Chapter 6, this volume)

Continued

Table 10.1. Continued.

Month and year	Incident or announcement	Comments/toxicological implications
	River Thames (UK) is severely polluted with plastic	Crabs, clams, over 100 species of fish and seals in the estuary are under threat due to ingestion of plastic and synthetic polymers
	Global migratory river fish populations decline by 76% in the past 50 years, with the greatest fall in Europe	Effects attributed to dams, overfishing, climate crisis and water pollution with plastic, sewage, fertilizers and pesticides
	Northern Europe radiation spike linked to reactor in Russia	Higher levels of ruthenium and caesium detected together with artificial radionuclides; there is no threshold level for radiation harm in humans (Chapter 8, this volume)
	Explosions at a nuclear power installation and a petrochemical facility in Iran	Radiation and petrochemical risks for workers and local residents are implied from other evidence (Chapter 8, this volume)
	Orange acid streams observed emanating near abandoned mine in Russia	An emerging ecological emergency requiring international vigilance
	Skin cancer mortality more than doubles in 50 years	Due to unprotected exposure to UV radiation (Chapter 8, this volume)
August 2020	International concerns over Mauritius crude oil spill from wrecked tanker	Extensive damage expected to pristine coral reefs; marine animals at severe risk (Chapter 5, this volume)
	Air pollution increases risk of death from COVID-19	Exacerbation of the well-known link between urban pollution and premature mortality of individuals with respiratory disorders (Chapters 3 and 10, this volume)
	Compensation over diesel emissions scandal expected	Use of defeat software by German car manufacturers to conceal actual nitrogen dioxide emissions; gaseous pollutants and particulates associated with increased incidence and severity of respiratory and cardiovascular disorders

Continued

Table 10.1. Continued.

Month and year	Incident or announcement	Comments/toxicological implications
	Genetically modified mosquitoes to be introduced in Florida as an alternative to insecticides	This approach is unprecedented and needs vigilance for unintended consequences affecting threatened species that feed on mosquitoes
	Dire prospects for native freshwater fish, with 22 species given less than 50% chance of survival in Australia	Attributed to climate change and invasion of alien fish, but pollution from farms and sewage may be a contributory factor
	Plans to destroy ancient woodland to create 40 new coalfields in India	Mining and combustion of coal is unequivocally associated with harm to human health and land degradation (Chapter 5, this volume)
	US proposals for oil and gas leasing plans in Alaskan wildlife refuge	Polar bears, caribou and other wildlife at risk from inevitable pollution
	Conglomerate admits delay in awarding compensation to individuals affected by herbicide	A total of 52,500 US claimants allege development of cancer due to exposure to glyphosate-based herbicides
September 2020	Alert over predictions of 'ecological disaster' in Kenyan Rift Valley lakes; deaths of over 300 elephants at waterholes in Central Africa	Wildlife at risk from algal blooms in alkaline and/or stagnant waters. Increased occurrence of algae attributed to climate change
	Electric cars unlikely to solve pollution problems	Particulates from tyres and road surfaces will still represent major contaminants; recycling used tyres also another issue; sourcing and disposal of Li, Cu and Ni in batteries will create additional pollution issues
	Decision to continue operations at Australian copper smelter refinery beyond 2022	Absence of any environmental risk assessment for local communities typifies general attitudes to impacts of industrial pollutants; smelting operations associated with increased sulfate particulate matter pollution

Continued

Table 10.1. Continued.

Month and year	Incident or announcement	Comments/toxicological implications
	Ozone pollution has increased in the northern hemisphere over the past 20 years	On its own or in combination with particulate matter, ozone increases risk of premature mortality from a variety of underlying diseases
	Insecticides to be sprayed in New York to eliminate Nile virus-infected mosquitoes	Human health risks include rashes, eye and nasal irritation as well as exacerbation of respiratory illnesses, for example, asthma
	US Administration weakens rules on toxic wastewater from coal plants	Effluents from impoundments may contain major and trace mineral elements that exceed regulatory guidelines for drinking water and ecological safety
	Escherichia coli contamination of lettuce in USA	Six deaths and 219 hospitalizations; attributed to manure from nearby livestock farms; coordination difficulties for regulatory agencies
	Inquest over Lewisham girl's death in 2013 'not about blame culture'	Refers to repeated asthma attacks caused by air pollution in London; episodes coincided with spikes in nitrogen dioxide; risk that no remedial action will be enforced
	World's wealthiest 1% cause double carbon dioxide emissions of poorest 50%	A sobering observation; carbon dioxide emissions inextricably linked with outputs of toxic gases and particulates in affluent societies
	Mining firms threaten endangered species in Hwange National Park (Zimbabwe)	Coal mining and smelting operations; noise and pollution concerns; air quality worse than in urban areas
	Toxic pesticides exported from the UK	Refers to paraquat and 1,3-dichloropropene; banned for EU use; implicated in neurodegenerative disorders and cancer, respectively
	All rivers and lakes in England polluted with sewage, fertilizers, pesticides, slurry and industrial chemicals	Toxic risks for aquatic species; proliferation of bacterial pathogens; other observations indicate microplastic pollution in River Thames
	Record low number of British butterflies baffles scientists	The widespread use of pesticides may be a contributory factor

Continued

Table 10.1. Continued.

Month and year	Incident or announcement	Comments/toxicological implications
October 2020	US electric automobile manufacturer as potential investor in Indonesia	Indonesia is a major producer of nickel for batteries; toxicity of nickel is well established; absence of mitigation measures in the statement is a major omission
	Indonesia mass protests over reductions of environmental safeguards	Concerns over deforestation add to health issues linked to nickel mining (see entry above)
	UK supermarkets are members of a group lobbying against the introduction of clean air zones in polluted cities	Vehicle emissions exacerbate underlying illnesses such as asthma, COPD and CVD, causing premature mortality
	Bottle-fed babies consume vast quantities of micro- and nano-microplastics every day	Pervasive contaminants of food and drinking water; toxicological issues under active review
	Polluted air killing half a million babies a year worldwide	Gaseous pollutants as well as particulates are implicated (Chapter 3, this volume)
	Pollution may have exacerbated mortality in the COVID-19 pandemic	Based on estimates published by researchers at Max Planck Institute (Germany); both COVID-19 and pollutants damage the respiratory and cardiovascular systems; effects predicted by D'Mello (2020) in *Introduction to Environmental Toxicology*
November 2020	Chernobyl-related concerns over commissioning a new nuclear power plant in Belarus	Safety issues raised by neighbouring Baltic states, with Lithuanian authorities issuing iodine tablets to residents near the Belarus border. Ionizing radiation is linked to incidence of thyroid cancer, based on Chernobyl evidence (Chapter 8, this volume)
	Of the most polluted cities listed by WHO, 16 out of 20 are in India and Pakistan	Due to ambient $PM_{2.5}$ pollution caused by agricultural burning and vehicle emissions; $PM_{2.5}$ associated with respiratory and cardiovascular disorders (Chapter 3, this volume)

Continued

Table 10.1. Continued.

Month and year	Incident or announcement	Comments/toxicological implications
	Two major US-based international companies 'dodging environmental responsibility' for electronic waste	Findings of Environment Audit Committee (UK); loss of precious metals; in addition, complex toxicological issues associated with other metals (Chapter 6, this volume)
	Delay to introduce Low Emission Zone regulations in London placed poor residents at risk from pollution	The poverty–pollution–health axis requires addressing in any strategy to 'level up' social inequalities (Chapters 3 and 10, this volume)
	European bison numbers are in recovery, but 31 other species are now extinct, including freshwater dolphins, fish and frogs	A familiar warning issued by the International Union for the Conservation of Nature and other agencies worldwide; habitat changes and chemical and sewage pollutants in rivers and lakes (Chapter 10, this volume)
December 2020	Radiation spike at nuclear reactor in Finland	'No danger' to public or environment; statement requires external verification (Chapter 8, this volume)
	Over 75% of polluted UK streets are in London	Nitrogen dioxide levels in parts of London significantly exceed EU limits; particulate matter, ozone and sulfur dioxide will add to human health risks and should be quantified (Chapter 3, this volume)
	More oil drilling in the Arctic permitted by Supreme Court in Norway	Environmental concerns dismissed despite persistent pollution issues in the region (Chapter 5, this volume)
	Air pollution in London contributed materially to death of asthmatic girl	Ruling by coroner at the conclusion of a landmark case; ambient nitrogen dioxide and particulate matter exceeded WHO limits (Chapter 3, this volume); failure to communicate health risks to mother
	Ultrafine aviation particulates affect lung function	Health risks for communities residing near airports. Ultrafine particles penetrate organs, tissues and subcellular components. VOCs and noise pollution add to adverse effects (Chapters 3 and 10, this volume)

Continued

179

Table 10.1. Continued.

Month and year	Incident or announcement	Comments/toxicological implications
	Indigenous communities subjected to fossil fuel pollutants in Oklahoma (USA) amid 'environmental genocide' claim	Contamination of air, aquifers, rivers and soil; multiple cases of asthma, CVD and industry-specific cancers reported; reflects environmental inequalities and deprivation for indigenous communities in Australia and Brazil (Chapters 5 and 10, this volume)
	Markedly high levels of toxic metals found in sharks in pristine waters around the Bahamas	Refers to about 12 heavy metals, including mercury; bioaccumulation via the trophic chain is a global issue (Chapter 6, this volume)
	UK government permits use of bee-killing neonicotinoid insecticide	A post-Brexit development; refers to 'limited and controlled' applications of thiamethoxam on sugarbeet; prohibited in the EU; toxicity and pesticide resistance in target species are overriding risks (Chapter 4, this volume)
January 2021	Ex-Michigan governor and former State colleagues charged over Flint water impurities	Refers to lead and other contaminants; reflects social inequalities across the globe; community leaders need to address toxic air pollution in cities
	Long-term effects of air pollution on cognitive skills	Detrimental effects can be detected up to 60 years later, adding to concerns presented in Chapter 3, this volume
	Major multinational conglomerate based in UK facing court action over oil spills in Nigeria	Land and groundwater contaminated with crude oil in the Niger delta; human health and ecological implications for the 'levelling up' strategy (Chapter 5, this volume)
	Domestic wood burning three times more polluting than cars	Due to $PM_{2.5}$ emissions (Chapter 5, this volume)
	Air pollution increases risk of infertility	Attributed to $PM_{2.5}$ emissions; based on study in China
	Shrimps producing fewer sperm in polluted waters	Sewage, legacy chemicals and climate change implicated in these effects
	UK moth populations in severe decline	Artificial light and chemical pollution are major drivers for sharp reduction in numbers

Continued

Table 10.1. Continued.

Month and year	Incident or announcement	Comments/toxicological implications
	Water companies discharged raw sewage into English rivers 400,000 times in 2020	In Environment Agency (England) report (Chapter 10, this volume)
	Maternal transfer of plastic particles to fetus	Based on a laboratory study
February 2021	Air pollution linked to increased dementia risk	Specifically implicating $PM_{2.5}$, based on data in Seattle (USA); (Chapter 10, this volume)
	Marine scientists discover 27,000 barrels of DDT discarded in the Pacific Ocean	The toxicity of DDT is now beyond question (Chapter 4, this volume)
	The rush towards electric vehicles may damage the environment	Pressure on resources and contamination of drinking water with lithium (Chapter 6, this volume)
	Endangered US rivers in peril	Multiple causes including coal ash, abandoned lead mines, farm waste and sewage pollution (Chapters 5, 6 and 10, this volume)
March 2021	Chernobyl radioactive hotspot in Ukraine forest	Drone detection (Chapter 8, this volume)
	New nuclear waste repository in exclusion zone in Ukraine	Raises issues over the 'clean energy' credentials of nuclear power generation worldwide (Chapter 8, this volume)
	Release of Fukushima nuclear plant wastewater into the Pacific Ocean	Contains tritium; symbolizes international disregard for the well-being of marine ecosystems (Chapter 8, this volume)
April 2021	Sri Lanka beach coated in crude oil	Associated with spill from damaged ship; mortality of turtles
	High mortality of manatees off Florida (USA) coast	Due to algal blooms, run-off from fertilizer plant and sewage pollution (Chapter 2, this volume)

Continued

Table 10.1. Continued.

Month and year	Incident or announcement	Comments/toxicological implications
	Nuclear reactions smouldering in Chernobyl compared to 'embers in a barbecue'	Result of fission reactions erupting within an inaccessible chamber in the ruins of the site (Chapter 8, this volume)
	'Alarming' levels of 'forever chemicals' in US mother's breast milk	Potential harm to infants associated with PFAS (Chapters 4 and 7, this volume)
	Women exposed to air pollution during pregnancy 'more likely to have children with asthma'	Ultrafine particles implicated in this effect (Chapters 3 and 10, this volume)
	Climate crisis creating anoxic conditions in lakes worldwide	Due to increased incidence of algal blooms (Chapter 2, this volume)
	World's soils under 'great pressure'	UN pollution report; due to accumulation of POPs and toxic metals (Chapters 4, 6 and 10, this volume)
	Communities in US Virgin Islands exposed to oil droplets from refinery	Refinery with history of environmental violations; concerns over safety of drinking water (Chapter 5, this volume)
May 2021	Antidepressants in water place crayfish in peril	Due to behavioural changes exposing crayfish to predators; pharmaceutical pollution is severely underestimated (Chapter 7, this volume)
	Plastic chemicals found in herring gull eggs	Relates to phthalates and transfer to offspring via contaminated eggs, potentially causing oxidative stress (Chapter 7, this volume)
	UK pig producers doubled use of antibiotics important for humans	Potential for development of antibiotic resistance and reducing efficacy for treatment of human diseases (Chapter 7, this volume)
	Children in poor countries exposed to electronic waste contaminants	WHO report on minors working at unregulated dumpsites (Chapter 6, this volume)

Continued

Table 10.1. Continued.

Month and year	Incident or announcement	Comments/toxicological implications
June 2021	Lead from gasoline still in London's air 20 years after ban	Estimated that 40% of lead in particulates is legacy contamination (Chapter 6, this volume)
	Concerns over radiation leaks at nuclear plant in China	Reflects similar incidents worldwide (Chapter 8, this volume)
	Significant oil slicks contaminate Corsica's coastal waters	Ships discharging crude oil into the sea (Chapter 5, this volume)
	COVID-19 may result in emergence of 'superfungus'	An example of co-adaptation of two pathogens (Chapter 9, this volume)
	'Forever chemicals' in cosmetics	Reflecting widespread use of PFAS linked to cancer, birth defects and endocrine disruption (Chapters 4 and 7, this volume)
	Air pollution exacerbates severity of COVID-19 cases	Based on findings in Detroit (USA); $PM_{2.5}$ and ethnicity implicated as factors (Chapter 10, this volume)
	Radon hotspots revealed by Public Health England	County of Somerset has elevated levels of the radioactive gas; higher in countryside than in large cities
	Dead fish on Florida (USA) shores amid 'red tide bloom'	Due to toxic algae (Chapter 2, this volume)
	Lawsuit against mining conglomerate	Relates to dam collapse in Brazil releasing mining waste into surface waters, including Atlantic Ocean (Chapter 6, this volume)
	Drugs turning trout into addicts	Various illicit pharmaceuticals regularly occur in rivers and streams (Chapter 7, this volume)
	Toxic metals accumulating in our bones	Attributed to lead released into air during production of smartphones, batteries, solar panels and wind turbines (Chapter 6, this volume)
	Migrating species in Asia and Pacific region are most vulnerable to plastic pollution	Impacts on river dolphins, Asian elephants and migrating birds highlighted in UN report (Chapter 10, this volume)
	'Evolutionary trap' for sea turtles exposed to plastic pollution	Ingestion of plastic polymers in evolved behaviour (Chapter 7, this volume)

Continued

Table 10.1. Continued.

Month and year	Incident or announcement	Comments/toxicological implications
	Mass mortality of fish in saltwater lagoon on Mediterranean coast of Spain	Attributed to fertilizer run-off from nearby farms; algal blooms also implicated (Chapter 2, this volume)
July 2021	LED streetlights detrimental to insects	Abundance of moth caterpillars reduced by LED street lamps (UK study)
	Gut microbiome determines susceptibility to 'Konzo' in children	Associated with differential release of HCN from cassava staple (Chapter 2, this volume)
	Emissions of particulates from California (USA) wildfires associated with preterm births	Based on work at Stanford University (USA) for the period 2006–2012 (Chapter 3, this volume)
	Transport noise linked to dementia risk	Attributed to release of stress hormones, immune disruption and inflammation (Chapter 10, this volume)
	French Polynesia: impact of decades of testing nuclear weapons	Toxic legacy of radionuclides emphasizes difficulties for clean-up at other sites, for example Chernobyl and Fukushima (Chapters 8 and 10, this volume)
August 2021	Oil spill spreads along Syrian coast	Satellite images increasingly used in environmental monitoring (Chapter 5, this volume)
September 2021	Fruit and vegetables contaminated with pesticides	Report of Pesticide Action Network (UK); produce contained traces of 122 different pesticides (Chapter 4, this volume)
October 2021	Alert on environmental triggers in carcinogenesis	Based on work on oesophageal cancer at the Sanger Institute (UK), implying different mechanisms in malignant initiation and development (Chapter 10, this volume)
	Bee colonies decimated across the USA	Attributed to disease and climate change; exposure to farm pesticides deserves consideration (Chapter 4, this volume)
	California oil spill renews efforts to prohibit offshore drilling	Severe wildlife impacts (Chapter 5, this volume)

Continued

Table 10.1 Continued.

Month and year	Incident or announcement	Comments/toxicological implications
	'Toxic tours' in Ecuador: oil spills	Designed to highlight ecological scandal caused by industrial contamination: turning disaster into tourism (Chapter 5, this volume)
	Clean environment should be incorporated in UN human rights charter	Proposal opposed by the USA and the UK
	Polluted water crisis in England (UK)	Water UK attribute high abstraction rates, pesticides, waste run-off from farms and sewage as underlying factors
February 2022	One in three Americans display detectable levels of toxic herbicide	Based on observations at George Washington University (USA); see also alert issued in September 2021

[a]COPD, chronic obstructive pulmonary disease; COVID-19, coronavirus disease; CVD, cardiovascular disease; DDT, dichlorodiphenyltrichloroethane; EU, European Union; LED, light-emitting diode; PAHs, polycyclic aromatic hydrocarbons; PFAS, perfluoroalkyl and polyfluoroalkyl substances; POPs, persistent organic pollutants; UN, United Nations; UV, ultraviolet; VOCs, volatile organic compounds; WHO, World Health Organization.

10.2   Distribution of Contaminants and Pollutants

10.2.1. Ambient air pollution emerged as a major stressor for human health in 1952, during the Great Smog episode, resulting in illness and death among residents in London. However, risks remain to the present time. Examination of Table 10.1 will show that as recently as December 2020, over 75% of polluted UK streets were in London. Other surveys consistently demonstrate that urban air pollution is a serious issue in cities around the world. New Delhi (India), for example, has been designated as the most polluted city on earth.

Factors contributing to ambient air pollution include:

- sources;
- chemical composition:
    - gaseous pollutants;
    - particulates;
    - persistent organic pollutants (POPs);
    - radionuclides;
- geographical and regional variations, for example in:
    - China;

- ○ North America;
- ○ Japan; and
- the impact of intensive agriculture.

The main pollutants in urban environments are those emitted in exhaust fumes, namely carbon monoxide, nitrogen dioxide, sulfur dioxide and particulates as well as ozone arising during photochemical smog formation. Despite early recognition of the adverse human health effects, air pollution has continued at an unprecedented pace in all industrialized countries.

In China, air pollution has been endemic for several decades as the country aggressively competes in international markets. A comprehensive analysis of the regional and temporal variations of urban air quality, published in 2020, focused on a wide range of gaseous and particulate pollutants. This risk assessment was based on national air quality ground observations for 300 cities in seven regions of China during the period 2014–2018. Relative to 2014 data, significant decreases in air pollutants occurred by 2018, particularly for emissions of carbon monoxide, nitrogen dioxide, sulfur dioxide and particulates. These improvements were attributed to rigorous application of internal control measures in China along with changes in meteorology. However, ozone levels increased during this period.

There were also regional variations, with high levels of pollution in the North China plains and in cities in Central and Western Xinjiang Province. Other observations indicated that the main sources of this pollution were associated with industrial activity and energy production in the Shandong Peninsula. In contrast, automobile exhaust emission constituted the major source of air pollution in the Beijing-Tianjia-Hebei region. These regional and point sources of air pollutants are replicated worldwide, particularly in the USA and Europe.

In particular settings, ambient air may be contaminated by POPs, particularly dioxins, dibenzofurans and polycyclic aromatic hydrocarbons (PAHs). Data appearing in 2018 indicated that in Beijing (China), higher atmospheric proportions of dibenzofurans relative to dioxins were associated with industrial activity and combustion processes, including municipal solid waste incineration. Levels of these POPs were lower in rural areas compared to industrial and residential sites in the city.

PAHs are also ubiquitous in ambient air as levels increase in response to the incomplete combustion of fuel, particularly coal, wood and petroleum products. Gas-particulate partitioning is an important factor in the dynamics of PAHs in the atmosphere. In addition, nitro-PAHs are formed directly from diesel and gasoline as well as via oxidants in the atmosphere. Although parent PAHs in ambient air are higher, there are increasing concerns over the higher relative toxicity of the nitro form.

Human responses to particulates depend on a variety of intrinsic factors relating to toxic air exposure. Particulates are a heterogeneous group of pollutants distinguished not only by dimensional classification into PM_{10}, $PM_{2.5}$ and ultrafine (nanoparticles), but also on physiochemical characteristics, whether organic or inorganic. In addition, radioactive particles may also occur in ambient air, depending on proximity to nuclear power installations or to accidental emissions.

Ambient distribution of particulates is, therefore, determined by mode of generation in different settings. In Canada, for example, the industrial contribution represented 64% of total emissions of particulates, as at 2016. In Utah (USA), ambient air particulates containing lead, copper, zinc and iron were detected in the vicinity of a steel mill. Similarly, air quality may be reduced by the presence of sulfate particulates near copper smelters. There are claims that approximately 90% of all south-west USA sulfate emissions in the 1960s were linked to activities at copper smelters in New Mexico, Arizona, Utah and Nevada (USA).

Results are now emerging on the atmospheric contamination in the aftermath of the Fukushima Daiichi nuclear plant accident. It appears that in excess of 99% of the radionuclides released comprised the highly volatile radionuclides of iodine and caesium. Although a substantial proportion of these emissions were dispersed over the Pacific Ocean, approximately 20% was deposited primarily over the highland regions of Japan. Over the past 8 years, high activity levels of the Fukushima-derived radiocaesium in surface air have persisted, with the soil surface providing a potential source of this isotope in the atmosphere.

Saharan dust events may also impact on atmospheric distribution of radionuclides. For example, in 2004, anthropogenic radionuclides consisting of caesium, uranium and plutonium were transported from the Sahara to the north-west Mediterranean Sea. Other evidence indicates that forest trees may serve as a sequestering system for the uptake of radionuclides contained in atmospheric emissions following accidents at nuclear power installations. For example, radioiodine may be incorporated into the xyloglucan fraction of trees. It is logical to speculate that wildfires such as those occurring regularly in Australia and California may redistribute any sequestered radionuclides into the atmosphere; although there is limited experimental evidence to support this notion, particularly for the release of radiocaesium.

In rural areas dominated by intensive agriculture, air quality is regularly compromised by ammonia emissions and particulates, adding to the human health risks caused by pesticide aerosols. The issue with safety of pesticides is dominated by food contamination, whereas inhalation of airborne sources might well be a major risk factor for occupational and residential exposures near farms.

- Name a common pesticide that might be present in aerosols.

10.2.2. It is widely acknowledged that the soil ecosystem contains a diverse array of contaminants as a direct result of anthropogenic interventions. There are also concerns and growing evidence that soils constitute a reservoir of volatile contaminants which may be recycled into the atmosphere. Consequently, it is important to determine the dynamics of the soil–ambient air interaction in considering the environmental fate of common pollutants.

Soil pollutants of toxicological significance include:

- agricultural chemicals;
- veterinary drugs;
- microbial pathogens;
- POPs:
  - Agent Orange;
- wood preservatives;
- toxic metals;
- effluents associated with shale oil and gas extraction; and
- radionuclides:
  - legacy issues.

Pollution of soils is inevitable, given the diverse range of agricultural and industrial activities across the globe. In arable farming, excessive applications of fertilizers and pesticides leads to retention in soils and contamination of watercourses from land run-off. Residues of antibiotics and other veterinary drugs may also contaminate soils on livestock farms. Microbial pathogens frequently occur in these soils.

Industrial activities lead to the deposition of a wide range of chemical residues in soil which may already be burdened by legacy contaminants. Data at a Superfund site in Florida (USA) are worthy of consideration in relation to POPs in industrial soil. On the basis of sample locations, dominance of specific congeners and derivation of total toxic equivalents, soil polychlorinated dibenzo-p-dioxins (PCDD)/ polychlorinated dibenzofurans (PCDF) status was 'reasonably' attributed to industrial activity at the contaminated site.

In addition, despite the passage of almost 50 years, dioxin levels remain high in soils and sediments contaminated with Agent Orange by the US Air Force in the Vietnam War during the period 1961–1971. Pollution was particularly severe in soils in and around the aircraft bases where the mixture of herbicides contaminated with dioxins was stored prior to deployment by aircraft.

Work in Canada indicates that pentachlorophenol and chromated copper arsenate, widely used as wood preservatives, regularly leach from treated surfaces to contaminate soil at commercial timber yards. Plants growing in the vicinity of these sites are able to translocate and sequester trace elements and organic pollutants in their aerial tissues. Consequently, these plants should also be categorized as potentially toxic, although there are favourable implications for phytoremediation of soils at contaminated sites. Analogous risks have been identified in the use of pentachlorophenol-treated timber poles at the National Wildlife Refuge site in Alaska (USA). Surface soil around these poles contained significant levels of the preservative, as well as 2,3,7,8-tetra-chlorodibenzo-p-dioxin (TCDD), which exceeded recognized limits for human health and ecological safety.

Other POPs, including PAHs and polychlorinated biphenyls (PCBs) may also contaminate soils, particularly in highly industrialized areas, as in the Dilovasi region of Turkey. Soils there were deemed to represent a potential risk to human health. Concerns have been expressed elsewhere that volatilization is a major factor in the transfer of POPs from soil to air. However, other factors, including congener composition, soil organic matter content, ambient temperature, local topography and long-range atmospheric transport may all modulate this partitioning effect.

In urban areas, the presence of toxic metals in the soil due to traffic emissions or from industrial sources may endanger human health, particularly children. A 2013 study of cadmium, nickel and zinc in surface soils near kindergartens and schools in Belgrade led to the conclusion of no danger to children from these pollutants. However, that investigation did not include risk assessments for other pervasive soil contaminants such as lead, mercury or POPs. In rural settings, the use of relatively high levels of copper sulfate as a growth promoter in the diets of growing pigs results in contamination of soils with this compound.

Of greater significance is the pollution caused by shale oil and gas extraction as well as by power generation by-products in routine operation or following accidental discharges. For example, heavy metal pollution around a coal-fired power station in China has recently been reported. The complex range including Pb, Cd, As, Hg, Cu and Cr in nearby soils and severe contamination of cabbages grown in that area is a cause for concern particularly as these metals may be accompanied by emissions of PAHs.

In the case of radionuclides, there is strong correlation between soil contamination and internal exposure of residents living in regions affected by nuclear accidents. Consumption of locally grown foods is considered to be the source of this exposure. In 1996, 10 years after the Chernobyl accident, Ukraine was divided into zones based on soil contamination of radionuclides.

The Fukushima accident caused serious radiocaesium deposits in soils in a range of terrestrial ecosystems. The dynamics of radiocaesium in relation to soil fractions, particularly clay minerals in surface layers are considered to modulate radionuclide behaviour in the environment, affecting mobility and bioavailability of the radionuclide. In both Chernobyl and Fukushima, radiocaesium dispersion was faster in forest soils than in grassland soils. There is, therefore, a strong perception that geochemical behaviour of radionuclides in terms of spatial and vertical distribution in surface soils constitutes an important factor in risk assessment for humans. It should also be recognized that, at least in the case of the Fukushima accident, a significant activity of radiocaesium still remains in the soil surface which may contribute to atmospheric recontamination of ambient air with this radionuclide.

It is clear, therefore, that soils worldwide carry a complex and significant burden of contaminants that may compromise both human health and biodiversity. In several cases, prospects for remediation are restricted by the persistence of contaminants in different soil types.

* What is a 'Superfund' site?

10.2.3. The claim that the Mediterranean Sea is in peril relates to extensive marine pollution, almost certainly by all 21 countries bordering this basin. European Union (EU) regulations are only enforceable in the northern half of the Mediterranean, with the southern half remaining subject to less stringent surveillance and compliance measures.

The range of contaminants in the Mediterranean Sea reflects worldwide misuse of surface waters in the environment. In virtually every geographical region and society, rivers and estuaries are perceived as conduits for waste disposal, while lakes, seas and oceans are used as reservoirs for harmful contaminants, left as a legacy for future generations to mitigate.

All the major classes of pollutants occur in the Mediterranean Sea, impacting both human health and biodiversity. However, it is also relevant to draw comparisons with other examples of aquatic pollution. Issues of particular significance relate to:

- sewage;
- organic chemicals:
    - perfluoroalkyl compounds;
    - PCBs;
    - pesticides;
    - crude oil;
    - PAHs;
- toxic metals;
- plastic debris of diverse grades;
- noise; and
- changes in agricultural practices.

Pollution of surface waters is widespread and seemingly out of control, even in developed countries. Thus, all rivers in the UK are polluted with plastic debris, with the Mersey containing microbeads, fibres and fragments in concentrations exceeding those in the Great Pacific Garbage Patch. All rivers and lakes in England are also polluted with sewage, fertilizers, pesticides, slurry and industrial chemicals. In 2020, a Scottish utility company pleaded guilty to a charge of polluting the Clyde with untreated sewage. The paltry fine imposed is unlikely to deter this and other companies from further breaches of environmental regulations.

Of the organic chemicals, the perfluoroalkyl residues in cetaceans have caused concerns as these may reflect a general picture of widespread pollution with these and other POPs. Perfluoroalkyls are a class of organohalogenated chemicals similar to other synthetic contaminants in terms of persistence, biomagnification and toxicity. Despite industrial reductions in production and use, these compounds still appear in high concentrations in the Mediterranean Sea. Stranded cetaceans in the western Mediterranean Sea during the period 2013–2018 were found to contain these compounds in the liver, with levels higher than in the muscle. Bottlenose dolphins had the highest liver concentrations of total perfluoroalkyls, followed by striped dolphins and sperm whales. Interspecies differences were attributed, at least in part, to habitat preferences of these cetaceans.

PCBs and pesticides of diverse chemistry are distributed widely in the Mediterranean Sea, extending, in particular, from the Spanish coastline towards the Adriatic in the vicinity of urban and industrial areas.

As might be expected, crude oil and associated compounds such as PAHs are significant pollutants in the Mediterranean Sea, with refineries, terminals and ports providing continuous sources of fuel contaminants. Accidental and deliberate discharges of crude oil are a dominant feature of pollution in different parts of the Mediterranean Sea. Crude oil discharge amounting to 13,000 t, associated with an incident in 2006 at the Jeyeh power plant in Lebanon, was a significant contributor to pollution. However, the preceding oil spill from the vessel MV *Haven* in 1991 off the coast of Genoa is judged to represent the fifth largest pollution incident of its kind since 1967.

A consistent feature of pollution in the Mediterranean Sea is the occurrence of a wide range of toxic metals, particularly lead, mercury, arsenic and cadmium linked to industrial and domestic waste sources. Although seawater concentrations of lead are in decline, correlating with changes in additive uses for gasoline, levels are still higher than those for the Atlantic and Pacific Oceans. Furthermore, lead contamination varies considerably, with lowest levels in protected habitats and highest in the vicinity of lead mining or processing sites on the Mediterranean coast.

Comprehensive surveys continue to emphasize extensive pollution with plastic in the Mediterranean Sea. Major contributors to 'hotspots' identified in June 2020 include Egypt, Turkey and Italy (Table 10.1). The main feature of plastic pollution is the diversity in the abundance and in the polymeric composition of the floating debris. The spatial distribution of polythene, polypropylene and polyamide are the predominant polymers, irrespective of distance from the coastline. Small fragments are more prevalent in coastal waters and correlate with close proximity to human habitation. Marine plastic debris also acts as vectors of potentially harmful microbes (Zhang *et al.*, 2022).

In 2020, the first record of 'plasticrusts' and 'pyroplastic' in the Tyrrhenian Sea in the Mediterranean was published. Plasticrusts arise through the force of sea waves on polyethylene litter impacting against intertidal rocks. Pyroplastic fragments are generated when polyethylene terephthalate debris is burned. These distinctions are not academic in view of reliable reports that UK plastic waste exported for recycling is instead burned at roadsides in Turkey with potential toxicological and environmental consequences (Table 10.1).

Marine noise is a particular issue in the Mediterranean Sea, associated with maritime traffic and industrial operations. In fish and invertebrates, such disturbances cause a wide range of adverse effects on communication, behavioural patterns, health status and even survival. The adverse effects are particularly significant for sonar-sensitive species such as sperm and Cuvier's beaked whales which appear to be deserting these waters.

Despite the foregoing, it should be noted that the disturbing extent of contamination in the Mediterranean Sea is replicated in other aquatic ecosystems across the globe. The Ganges in India and the Santiago River in Mexico contain a diverse burden of contaminants, including tannery effluents, heavy metals, sewage, animal wastes as well as organochlorine pollutants at levels well in excess of acceptable limits. In China, the Yangtze and Pearl rivers are contaminated with antibiotics at concentrations below medical doses, but nevertheless representing risks for the potential development of resistance in human pathogens.

Changes in agricultural practices, including certain conservation measures have inadvertently resulted in the development of harmful algal blooms in western Lake Erie (Canada). This effect is attributed to increased agricultural phosphorus and nitrogen loading from tributaries draining into the lake. Farm pollution is responsible to a large extent for the destruction of coral reefs around the world.

In conclusion, the statement that the Mediterranean Sea is 'dying' is fully justified by the scientific evidence. However, it is important to recognize that rivers around the globe are primary routes for marine pollution, emphasizing the imperative and urgency for regulatory measures at local and international levels.

- Can you define 'cetaceans'?
- Outline the toxicological hazards of burning plastic.

10.2.4. Biomonitoring of surface waters is generally conducted by determining bacterial load, particularly in rivers, well water and coastal lagoons. Several genera can be used as potential diagnostic tools for this purpose. However, in practice water quality is assessed by the occurrence and abundance of 'indicator' species of microbes, such as:

- *Campylobacter jejuni* and *Campylobacter coli*;
- *Escherichia coli*;

- *Enterococcus*; and
- severe acute respiratory syndrome coronavirus 2 (SARS-CoV-2) RNA.

Examination of Table 10.1 shows that, for example in England, all rivers failed quality tests due to the presence of unacceptably high contamination with sewage, agricultural run-off and animal slurry from nearby farms. This pollution invariably results in microbial contamination of surface waters and drinking supplies.

According to a 2018 study, *C. jejuni* and *C. coli* represent the most frequently recorded bacterial enteropathogens in the EU and other developed countries. Although these bacteria occur in poultry, adaptation mechanisms permit survival of these microorganisms in lakes and well water; consequently, microbial transmission to humans occurs via contaminated drinking water.

The presence of *E. coli* in surface waters is also frequently reported by monitoring agencies. The association with sewage contamination means that this bacterium is also used as an indicator of both water quality and food safety in irrigated agricultural/horticultural land. Examination of Table 10.1 will show that in 2020 an alert was issued with regard to *E. coli* contamination of lettuce in the USA. Six deaths and 219 hospitalizations were reported and attributed to manure from nearby livestock farms contaminating irrigation water. Separately, other observers comment on a general and regular increase in food-borne illnesses in the USA due to consumption of fresh produce, again associated with bacterial contamination of irrigation water.

Monitoring of the quality of coastal bathing waters and beaches in Spain with respect to contamination with faecal bacteria including *E. coli* and *Enterococcus* has revealed spatial variations. It appears that sandy beaches harbour higher concentrations of these bacteria compared to gravel coastlines. In addition, as might be expected, urban coasts are more polluted with bacteria than semi-urban or natural beaches.

Water quality is a particular issue in parts of Africa and Asia, with high incidence of gastrointestinal and other diseases. For example, in Nigeria, high heterotrophic bacteria, faecal coliforms and enterococci counts, above permissible limits for drinking or recreational waters, in the River Sokoto have compromised water quality and safety. Microbial contamination of water in many developing countries regularly exceeds permissible limits established by monitoring authorities such as the WHO and the EPA in the USA.

Emerging evidence points to the presence of SARS-CoV-2 RNA in sewage from coronavirus disease (COVID-19) cases, for example those identified in long-term care facilities (Lee *et al.*, 2021). The detection of this RNA in stool samples has been reported in 40–50% of such cases associated with diarrhoea, an early manifestation of COVID-19 infection. In view of widespread pollution of surface waters with untreated sewage, these observations further amplify ongoing concerns over safety of rivers and coastal amenities.

The question arises as to whether climate change is likely to exacerbate microbial contamination of surface waters. There is, clearly, a need to maintain a watching brief as precipitation and environmental temperatures rise.

- What are 'heterotrophic' bacteria?
- Can you name a waterborne bacterial disease?

10.2.5. It is axiomatic that pollution should adversely impact on safety of food, given the ubiquity of toxic compounds in different environmental compartments. On the basis of comprehensive analysis, D'Mello (2003) highlighted widespread and continuing environmental health concerns over food safety. It was predicted then that, despite the best efforts of established and new statutory agencies, food and water contamination would remain a significant risk for public health. Since 2003, the emphasis on food safety has been increased by emerging evidence concerning:

- diversity of contaminants;
- global scale of food safety issues;
- microbial contaminants and toxins;
- POPs in:
    - seafood;
    - meat;
    - eggs;
- toxic metals;
- radionuclides; and
- regulatory updates.

Microbiological contamination occurs more frequently in dairy and poultry products than chemical pollutants. *Listeria monocytogenes*, *Staphylococcus aureus*, *Salmonella* and pathogenic forms of *Escherichia coli* occur regularly in these products, requiring careful handling by commercial processors, distributors and consumers. Of the microbial toxins, aflatoxin M_1 continues to occur regularly in milk products, linked to the use of contaminated animal feed (Chapter 2, this volume).

Monitoring of dioxins and dioxin-like PCBs in foods and animal feeds has continued on a global scale, consistent with advances in analytical methodologies. Investigations in Colombia in 2016 typify widespread concerns over the occurrence of these POPs in local foods. Predictably, fish oils and shrimp present particular issues in this respect, followed by butter and soybean oil, with some values occurring above EU regulatory safety values. A subsequent survey in Japan indicated that average intakes of PCDD/PCDFs and dioxin-like PCBs were maximal from fish and shellfish followed by meat and eggs. However, consumption was well below tolerable daily intakes set for the population.

A disturbing observation has been the occurrence of PCDD/PCDFs and dioxin-like PCBs in free-range eggs obtained from a farming area near to an illegal waste-incineration site in Italy. Contamination in eggs regularly contravened action levels set by the EU, but tolerable weekly intakes of these POPs were lower than limits set by the EU Scientific Committee on Food. Nevertheless, it is important to consider risks from meat and milk produced on other farms in that vicinity.

It is reassuring that a watching brief relating to PCDD/PCDFs and dioxin-like PCBs is maintained by food safety agencies such as the European Food Safety Authority (EFSA). Nevertheless, it is salutary to emphasize the recalcitrant and long-term impact of POPs in foods. For example, despite the passage of almost 50 years, dioxin levels remain high in fish and shrimp from surface waters contaminated with Agent Orange by the US Air Force in the Vietnam War during the period 1961–1971.

Worldwide surveillance of pesticide residues in foods continues, as both legacy and emerging chemicals are linked to human health disorders. The export of paraquat and 1,3-dichloropropene by the UK has evoked alarm since these pesticides are banned for EU use due to associations with neurodegenerative disorders and cancer, respectively (Table 10.1). Although dichlorodiphenyltrichloroethane (DDT) use has been prohibited by law in the EU and the USA, surveillance in 2017 indicates relatively high levels in meat and breast milk in Bangladesh. In China, organophosphate, carbamate, pyrethroid and triazine residues were found in 42% of 12 samples of vegetables and fruit, with highest incidence of contamination determined in leafy vegetables.

Lead surveillance is ongoing worldwide, particularly in the USA, following the water contamination incident in Flint in 2014. Also in the USA, there are concerns over lead exposure linked to non-paint sources, particularly candy imported from Mexico. This 2017 investigation was instigated by several reports of childhood lead poisoning in California that were linked to candy and other food items.

A survey in Pakistan highlighted the wide distribution of lead, cadmium, nickel and arsenic in common plant-based foods to levels toxic for humans. There are also particular risks for residents on the periphery of major industries. For example, traditional foods of indigenous communities in the Bigstone Cree Nation in the vicinity of the Athabasca oil sands were contaminated with mercury, methylmercury and selenium. Overall, results indicated substantial variations in the toxic metal profile across a comprehensive range of traditional foods.

It is well established that seafoods contribute substantially to the body burden of mercury particularly in coastal communities reliant on these items. A 2020 study confirmed that fish/seafood consumption is reflected in biomarkers based on blood levels of total mercury and methyl mercury in conjunction with urinary mercury concentrations. In addition, however, it was concluded that consumption of rice, vegetables or alcohol may also contribute to levels of these biomarkers, especially among non-seafood consumers.

Compared to previous data reviewed in 2003, there is greater optimism regarding radionuclides in foods, following accidental contamination in Chernobyl and Fukushima. For example, a 2018 study indicated that most Japanese foods contained low but detectable levels of radiocaesium. However, it should be recalled that there is no threshold level for radionuclide toxicity and that atmospheric pollution may be a more important route of exposure to artificial radionuclides.

In 2019, it was concluded that consumption of seafood available in the Korean Peninsula was associated with annual effective doses of artificial radionuclides that were insignificant, relative to natural sources.

At about the same time a case report relating to radionuclides in reindeer meat in Sweden was also published. Declining environmental contamination of radiocaesium in the wake of the Chernobyl accident in 1986 has been accompanied by reductions in levels of radiocaesium in reindeer meat and few carcasses are now rejected for human consumption. In Serbia, it has been suggested that there are no radiological hazards associated with seafood but frequent consumption of this item may compromise health due to mercury and lead contamination.

There are clearly justifiable reasons for continued surveillance of foods for biological and chemical contaminants arising naturally or from anthropogenic activities. Although food safety is a global issue, it is important that in post-Brexit UK, for example, standards are maintained to at least EU guidelines.

- What are the main food safety issues in your country?

10.3 Human Health Emergency

10.3.1. Current priorities in evaluation of health risks associated with ambient air pollution in urban environments vary considerably in different countries. It is widely accepted that Mexico City is one of the most polluted locations in the world. The severe ambient air pollution problem there is attributed to:

- geographical factors;
- altitude;
- local topography;
- unpredictable meteorological events;
- demographic factors;
- exponential increase in the number of automobiles;
- socio-economic disparities that exacerbate the health issues for residents;
- the nature of pollutants; and
- novel aspects of morbidity.

Of all ambient air pollutants, $PM_{2.5}$ emission is the factor of particular concern in Mexico City, originating from diverse sources dependent on variations in spatial as well as temporal distribution. Predictive models based on geographical information systems and land-use regression criteria have revealed areas of particularly high $PM_{2.5}$ concentrations (up to 109 $\mu g\ m^{-3}$).

Novel aspects of morbidity and interactions have been associated with these levels of contamination. For example, epidemiological observations suggest increased risk of tuberculosis development in response to air pollution, including $PM_{2.5}$. The results of a 2018 study with healthy volunteers residing in Mexico City, suggest that inhaled $PM_{2.5}$ as well as other pollutants may interact with microbial pathogens in alveolar macrophages to increase risks for the induction of tuberculosis. Other observations published in 2020 indicate that ambient $PM_{2.5}$ may affect the prevalence of childhood acute respiratory syndrome in Mexico City. Experimental evidence suggests that inhalation of $PM_{2.5}$ from Mexico City may initiate allergic asthma *de novo*.

The effects of ambient air $PM_{2.5}$ on respiratory disorders in Mexico City are replicated worldwide in London, Los Angeles, New Delhi and other densely populated and industrialized conurbations. COPD is a global public health issue, affecting more than 300 million individuals and it accounted for an estimated 3 million fatalities in 2015. In China, COPD cases almost doubled during the period 1990–2013. There is compelling evidence in European and US investigations that ambient $PM_{2.5}$ is associated with incidence and exacerbation of COPD.

Work in China has lagged behind, with monitoring of atmospheric $PM_{2.5}$ not commencing until 2013 (Tian *et al.*, 2018). Emerging evidence, however, confirms that even short-term exposure to ambient fine particulates (and other pollutants) in Beijing was associated with both out-patient and in-patient hospital visits for COPD treatment. In a cross-sectional study reported in 2020, it was concluded that long-term exposure to both $PM_{2.5}$ and nitrogen dioxide in Shanghai (China) significantly impaired pulmonary function in adults. Furthermore, the effects were modified by gender, age, obesity and tobacco-smoking history.

These studies offer the opportunity to identify vulnerable individuals in populations exposed to ambient air pollutants. However, other investigators in Barcelona (Spain) cautioned that emergency visits and hospitalization for exacerbation of underlying respiratory disease such as asthma should take into account ambient temperatures in addition to pollution variables.

It has been estimated that 75% of the population in Europe live in urban areas. It follows that ambient air pollution in major European cities is associated with severe health impacts at personal and community levels. The ongoing air pollution crisis in London has been designated as 'lethal and illegal', and with justification. In London, continuing concerns include the possible effects of noise and air pollution on:

- incidence of dementia;
- risks of preterm birth;
- stillbirth; and
- correlations with cardiovascular and respiratory hospital admissions.

In London, long-term exposure to road traffic noise has caused annoyance, elevated blood pressure, cardiovascular disease and even mortality among residents, as in other urban settings. Other emerging concerns in London relate to interactions between nitrogen dioxide and $PM_{2.5}$ on mental health problems in children. In addition, socio-economic and ethnic inequalities associated with exposure to pollution are being addressed, presenting implications for other communities with diverse populations. The landmark developments following the death of a severely asthmatic child due to consecutive spikes in ambient nitrogen dioxide levels serves as an example of ethnic inequalities in London.

In London, as elsewhere, the association between ambient air pollution and incidence of ischaemic and haemorrhagic stroke is under regular review. Of considerable concern, to be elaborated later in this chapter, is the role of air pollutants in the induction of cancer. The IARC has designated these pollutants as Group 1 carcinogens.

It is clear, therefore, that ambient particulates are strongly implicated in the incidence of morbidity in Mexico City, with unique pathological interactions. Similar effects are replicated in other polluted conurbations worldwide. However, the role of oxides of nitrogen, sulfur and ozone in the exacerbation of respiratory and cardiovascular diseases should be factored into future investigative models.

- Briefly define 'tuberculosis'.
- What is 'ischaemic stroke'?

10.3.2. This question offers an opportunity to explore complex interactions in a challenging, albeit evolving issue concerning pollutant toxicology. Respiratory disorders including asthma, chronic obstructive pulmonary disease (COPD) and cystic fibrosis are the result of multidimensional interactions involving a complex array of risk factors. However, despite this knowledge, pathogen–pollutant interactions in the exacerbation of symptoms in lung pathology remain largely unexplored.

An insight into these interactions emerged in 2018 in a study on the impact of ambient air pollution on asthma outcomes in schoolchildren residing in urban Agra (India). A significant correlation between traffic-derived pollutants and respiratory infections and the development of clinical exacerbations in these asthmatic children was noted. This evidence is indicative of a pathogen–pollutant interaction, but conclusive proof remains elusive. Furthermore, the components of air pollution in Agra were not identified.

In the case of individuals with underlying COPD, there is limited evidence that higher levels of ambient nitrogen oxides and viral infections may be associated with prolonged exacerbations in these patients (Pfeffer *et al.*, 2019). This study was deemed to support the concept that the toxicological and clinical manifestations in COPD were exacerbated by ambient air nitrogen oxides and viral infections. An increased risk of exacerbations of probable viral aetiology was consistently observed 2–4 days after exposure to elevated nitrogen oxides in ambient air. In support of these tentative conclusions, Pfeffer *et al.* (2019) cite other evidence that nitrogen oxides increase expression of intercellular adhesion molecule 1, the primary entry receptor for human rhinoviruses and other respiratory pathogens.

Arguably, the clearest example of pathogen–pollutant synergy is illustrated by exacerbations in cystic fibrosis, a lung disease. Cystic fibrosis is an incurable autosomal recessive disorder caused by mutations in a gene that encodes a transmembrane system for transport of chloride and bicarbonate anions. This condition is characterized by pulmonary inflammation and infection resulting in significant morbidity and premature death.

The risk factors for cystic fibrosis, as presented in Fig. 10.1, include:

* genetic predisposition;
* ambient temperature;
* ambient air pollutants;
* microbial colonization of lungs:
  ○ *Pseudomonas aeruginosa*;
  ○ *Aspergillus fumigatus*:
    – direct impacts;
    – role of mycotoxins;
  ○ methicillin-resistant *Staphylococcus aureus* (MRSA) acquisition; and
  ○ viral infections.

The dynamics in cystic fibrosis should be considered in the light of limited evidence indicating an element of interaction between two variables: (i) pathogens; and (ii) ambient air pollutants.

Ambient air particulates, nitrogen dioxide and ozone can act as triggers in the exacerbation of cystic fibrosis, with increased need for oral or intravenous antibiotics on the day of exposure. Co-colonization of *P. aeruginosa* and *A.fumigatus* in cystic fibrosis tends to be high in the lungs of patients with cystic fibrosis. This prevalence is associated with exacerbation of cystic fibrosis.

In experimental bioassay models, exposure to ambient air pollution aggravates *Pseudomonas* infection in cystic fibrosis. It is entirely possible that mycotoxins produced by *A. fumigatus* may intensify clinical manifestations in this syndrome. Other observations imply

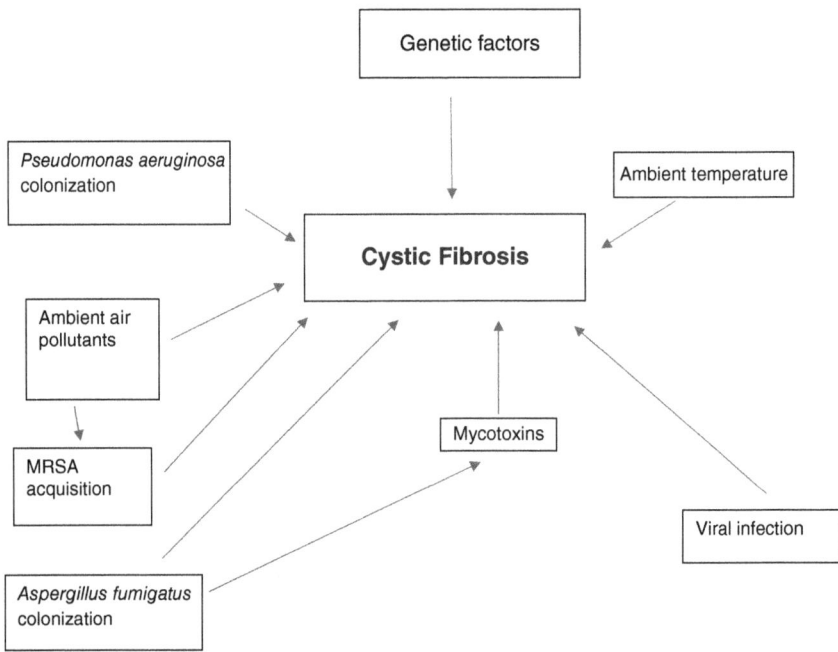

Fig. 10.1. Risk factors for cystic fibrosis: an instructional model for a debilitating disorder. Although each component of this model is shown separately, the thesis being tested in several ongoing investigations is that one or more of these factors may act synergistically with ambient air pollutants to exacerbate clinical manifestations and eventual outcomes in this disorder. Similar interactions may operate in other respiratory syndromes, including asthma and chronic obstructive pulmonary disease (COPD). See also Caverly *et al.* (2022) for updates on the microbiome in respiratory disease. MRSA, methicillin-resistant *Staphylococcus aureus*.

- Based on your lecture notes and supplementary reading, can you add further variables to this model?
- Have you thought about the possible impact of the COVID-19 pathogen?

that fine particulates in ambient air may act as a risk factor for initial acquisition of *P. aeruginosa* and MRSA in young children with cystic fibrosis.

Viral respiratory tract infections, for example, rhinovirus, influenza, adenovirus and coronavirus may also operate as triggers of exacerbation in cystic fibrosis. There is clear evidence that ambient air pollution and viral infections can combine to adversely affect the respiratory system. Consequently, such synergistic interactions may also exacerbate manifestations in cystic fibrosis. A logical question is whether COVID-19 infection may prejudice or prolong adverse effects in cystic fibrosis patients living in air-polluted cities.

There is limited physiological evidence for pathogen–pollutant inter-actions in pulmonary syndromes. For example, ambient nitrogen oxides increase expression of intercellular adhesion molecule 1, the primary entry receptor for human rhinoviruses and other respiratory pathogens.

In addition, it is known that the adverse effects of ambient ozone are mediated principally via products of its reactions with unsaturated components of alveolar surfactants and associated phospholipids. The primary ozonide products engage in further reactions to precipitate lipid peroxidation. It is conceivable that the resulting tissue damage may facilitate interactions with pathogens, particularly in individuals already or genetically compromised by respiratory disorders such as asthma, COPD and cystic fibrosis.

- Explain the meaning of 'autosomal recessive disorder'.
- What are 'rhinoviruses'?

10.3.3. Gene–environmental interactions are increasingly being in-voked to explain adverse human health responses to common pol-lutants. For example, antioxidant genes and susceptibility to air pollution have been considered for respiratory and cardiovascular health (Fuertes *et al.*, 2020). An intriguing question is whether pollu-tants and genetic predisposition interact in a synergistic pathway of mechanisms to accelerate onset of disease or to exacerbate eventual outcomes and thereby increase mortality (Fig. 10.2).

It is conceivable there might be genetic variations in oxidative stress and inflammation pathways and that pollutants can interact with these mechanisms, to accelerate onset and exacerbate clinical manifestations in conditions such as:

- cardiovascular disease:
  - genetic markers;
- asthma;
- COPD;
- cystic fibrosis;
- diabetes;
- neurodegeneration;
- cancer; and
- COVID-19:
  - ethnicity as a potential risk factor.

It is well established that cardiovascular disease is the result of a wide range of risk factors, including lifestyle choices, age, gender and

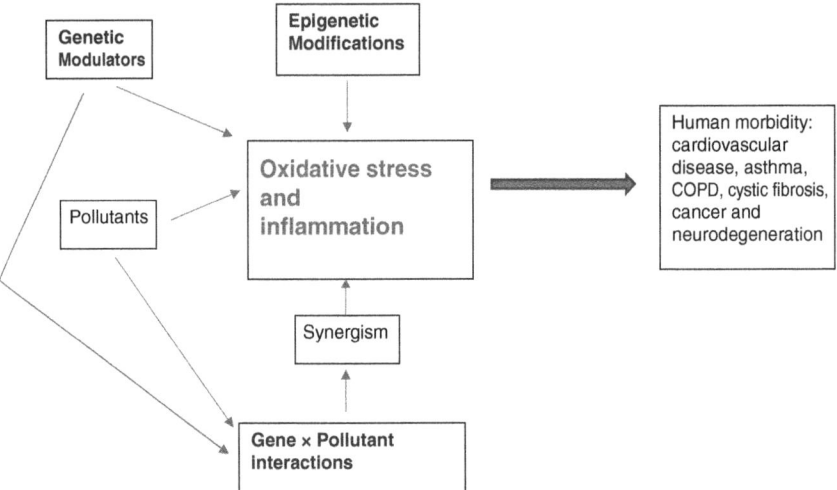

Fig. 10.2. A speculative model designed to illustrate the possible synergistic dynamics of gene–pollutant interactions in human morbidity (see also Melbourne *et al.*, 2022). It is widely recognized that major disorders such as cardiovascular disease, asthma, COPD, cystic fibrosis, cancer and neurodegeneration are determined, to a significant extent, by genetic predisposition. Exposure to pollutants is also a risk factor for these conditions. Both genetic modulators and pollutants may act independently. However, in addition, the possibility exists for synergism between these two variables to exacerbate onset and severity of clinical manifestations and to delay or prevent recovery. The impact of epigenetic modifications is increasingly emphasized in several mechanistic proposals and students should take special note of emerging advances in this aspect of environmental health disorders.

- Can you outline the major forms of cardiovascular disease manifestations?
- What are epigenetic modifications?
- How do these modifications differ from mutations?

inherited gene variants. Specifically, an international team of experts has identified the existence of a comprehensive range of genetic markers that increase cardiac disease risks and susceptibility in particular individuals. However, it is now argued that traditional risk factors merely account for a fraction of cardiovascular disease cases.

It is therefore conceivable that an environmental dimension in the form of exposure to pollutants should be considered as an additional risk factor in conjunction with genetic predisposition acting to exacerbate manifestations and outcomes via a synergistic mechanism. This hypothesis is now the subject of ongoing research. For example, preliminary data indicate that ambient air pollutants and genetic variants in inflammatory markers may interact and modify risks of non-fatal myocardial infarction. On the basis of a comprehensive evaluation,

Fuertes *et al.* (2020) concluded that evidence is limited, yet 'suggestive', for the interaction between antioxidant genes and air pollution in relation to cardiovascular health.

There is considerably greater agreement on genotype–pollutant interactions in the incidence of asthma. Thus, researchers in China concluded that polymorphisms of multiple genes and ambient pollutants including sulfur dioxide, nitrogen dioxide and particulates are risk factors for asthma.

Other observations, based on advances in epigenetics, provide evidence that DNA methylation may act as a mediator in gene–air pollution interaction. It was concluded that such environmental exposures were associated with higher risks for the development of asthma, particularly in the presence of genetic risk factors.

Additional insight is provided by evidence of a significant association between ambient air pollution and reduced lung function in minorities with asthma. Ethnicity is an indicator of genetic variation which also demonstrates striking disparities for minorities living with asthma in polluted urban environments.

The gene–pollutant interaction has also been explored in the incidence and development of diabetes. The emerging consensus appears to indicate that individuals at higher genetic risk for type 2 diabetes may be more susceptible to ambient particulates due to alterations in insulin sensitivity. However, this mode of action requires confirmation.

Initial concepts in gene–pollutant interactions emerged in aspects of insecticide toxicity following occupational exposure. For example, organophosphate compounds are widely used to control ectoparasites in sheep. However, a significant number of sheep farmers attribute chronic ill-effects to repeated exposure to these insecticides, while others are asymptomatic in this respect. It is possible that genetic differences in susceptibility may account for variations in adverse responses to these and, indeed, other pesticides. Current investigations indicate that the metabolic disposal of organophosphates in humans occurs principally via the paraoxonase enzyme which is associated with gene polymorphisms.

The gene–pollutant interaction may operate in the incidence and development of well-defined neurological disorders particularly for those conditions that are associated with a consistent element of heritable traits. In this respect, it has been proposed that the apolipoprotein E4 allele, the most prevalent genetic risk for Alzheimer's disease, exerts a key role in the response to air pollution constituents, $PM_{2.5}$ and nanoparticles.

It has been suggested that the majority of Parkinson's disease cases are likely to be the result of different combinations of environmental exposures and genetic susceptibility. In model studies, rotenone and paraquat have been shown to interact with several of the genes linked to Parkinson's disease, including for example, α-synuclein.

The development of ovarian and breast cancers has provided a significant and relevant model to test the gene–pollution interaction, given the ubiquitous distribution of endocrine disruptors in the environment. For example, polymorphisms of xenobiotic-metabolizing genes have been implicated in the aetiology of epithelial ovarian cancer following exposure to organochlorine pesticides. Other observations indicate that genetic variants can modify the risks of exposures to POPs in breast cancer incidence. In addition, there are concerns that women with higher breast cancer risks, based on family history, may be more susceptible to endocrine-disrupting pollutants.

In conclusion, despite the foregoing evidence, it is still premature to assess whether gene–pollutant interactions act in an additive or synergistic mode in determining outcomes in different manifestations of human morbidity. This issue is of immense significance, particularly for individuals living in polluted urban environments. Thus, it is emerging that ethnicity (and, therefore, genetics) is a significant factor in determining the response to COVID-19 infection, in terms of both severity of the disease and incidence of mortality. The superimposition of ambient air pollution in exacerbating disease outcomes for ethnic patients residing near congested roads and junctions has yet to be resolved. It is imperative that this issue is addressed through scientific analysis and not confounded by political expedience.

- What is your opinion on the effects of ethnicity in environmental health disorders?

10.3.4. It is consistently maintained that pollutants can contribute directly to the incidence of obesity in humans. However, it is of immense importance in modern society to know whether obesity induced by other factors such as genetic predisposition or lifestyle choices affect responses to common pollutants. It was estimated in 2014 that global incidence of obesity had nearly doubled since 1980. A high body mass index (BMI) is now regarded as a highly significant contributor to overall disease burden across the world.

The pathology of obesity is associated with increased systemic oxidative stress and inflammation. Exposure to ambient pollutants, including PM$_{2.5}$, also involves these mechanisms and, consequently, there is a common basis for additive or, more disturbingly, synergistic interactions that would lead to early onset or exacerbation of final outcomes.

Figure 10.3 depicts the emerging dynamics and dilemmas in obesity–pollutant interactions. The principal issues under review are:

- genetic predisposition;
- dynamics of adipose tissue:
  - toxic pool concept;
- associated morbidity:
  - diabetes;
  - asthma;
  - cardiovascular disease;
  - carcinogenesis;
  - endocrine disruption;
  - developmental impairments;
  - immunosuppression;
- lifestyle choices:
  - obesogenic diets;
  - sedentary regime; and
- evidence of exacerbations.

Interest in the role of adipose tissue has led to the development of the concept of chronic internal exposure as an additional risk factor for the toxicity of lipid-soluble contaminants such as POPs. It is envisaged that, following entry into the body, POPs are transported in the bloodstream to precipitate adverse effects directly at target sites. Absorbed POPs may also be diverted to a 'toxic pool' in adipocytes and stored there until, in women for example, extra demands of pregnancy or lactation cause mobilization of body fat to serve as a source of energy for the mother, fetus or suckling baby. The circulating levels of POPs so released may then act as internal sources of contaminants during gestation and subsequent lactation, exemplifying the concept of 'developmental toxicity'.

Obesity is also regarded as an independent risk factor by virtue of its direct effects on inflammation and other pathways for the induction of conditions such as diabetes and cancer. Whether POPs and obesity act additively or synergistically to influence onset or final outcomes is a critical question. This issue has been addressed in a National Health and Nutrition Examination Survey for a sub-population in the USA.

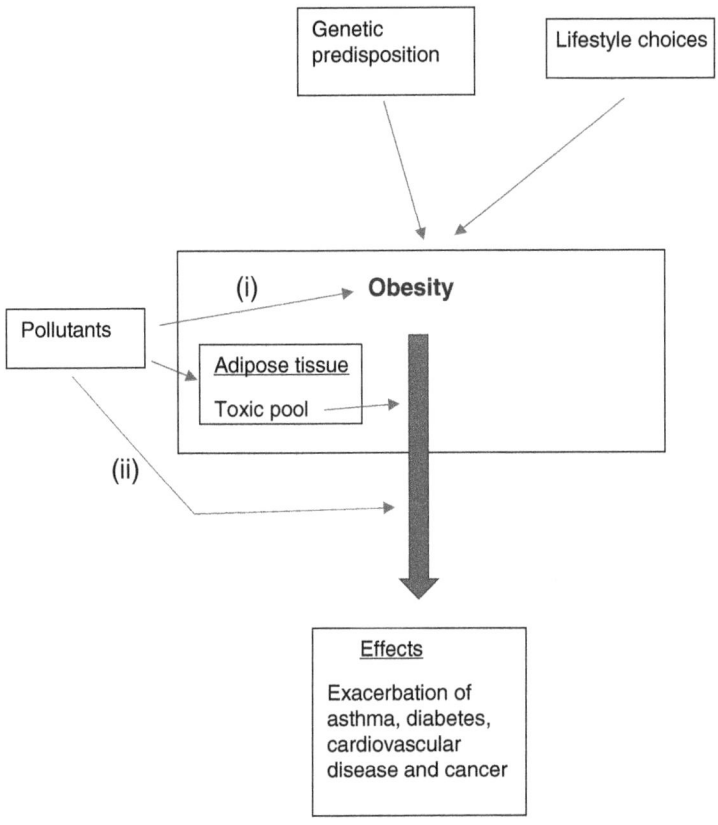

Fig. 10.3. Obesity–pollutant dynamics in humans: an illustrative model. The accelerating pace of obesity, particularly among children, is the subject of considerable concern in the medical profession. Although there are clear genetic determinants, it is consistently emphasized that lifestyle factors are major contributors to this disorder. Obesity may cause morbidity by direct mechanisms. In addition, pollutant exposures via inhalation and/or food intake are also associated with the induction of obesity and exacerbation of adverse effects indicated at (i) and (ii), respectively. Furthermore, circulating pollutants may accrete in adipose tissue to be mobilized subsequently in times of extra energy demands caused by pregnancy or lactation. The pollutants so released may adversely impact the developing fetus or the nursing offspring. These proposed metabolic transactions exemplify the concept of 'chronic internal exposure', particularly but not entirely, in relation to persistent organic pollutants (POPs).

- Can you name one contaminant in the 'toxic pool'?
- What is the clinical definition of obesity?
- Can you identify the main component of adipose tissue?

Emerging data indicated that the correlation between exposure to POPs and diabetes was greater among obese compared to lean individuals. The POPs under investigation included PCDDs, non-dioxin-like PCBs and organochlorine pesticides.

Obesity–pollutant interactions have also been postulated for metabolic and physiological conditions caused by other environmental exposures, for example, toxic air. There is evidence of marginal positive associations between ambient $PM_{2.5}$ and markers of systemic inflammation, with particular impacts for individuals with obesity as well as diabetes and hypertension. Positive correlations between ambient $PM_{2.5}$ and interleukin-6 production and white blood cell counts have been observed in COPD patients in a 2020 study, with stronger effects in individuals with diabetes, obesity and hypertension.

In the case of oxidative stress and cardiovascular risk, obesity, diabetes and cigarette smoking may interact in determining the outcomes of exposures to pollution. In particular, there is accumulating epidemiological evidence to indicate that obesity may increase susceptibility to the adverse cardiovascular health effects of ambient $PM_{2.5}$. Obesity also exacerbates the association of long-term air pollution with blood pressure and hypertension in children, based on observations in China.

Developmental toxicology continues to form a significant criterion in risk assessments for the principal contaminants impacting on human health. An important study based on modelled data indicates that prenatal exposure to ambient $PM_{2.5}$ in traffic-related pollution may result in reduced birthweights of progeny which was particularly associated with maternal obesity prior to pregnancy.

Obesity exerts critical impacts on the health of children postnatally as well and the problem continues unabated despite strong recommendations of remedial measures worldwide. Epidemiological observations in China confirmed that obesity amplified the association of long-term air pollution exposure with blood pressure and hypertension impairments in children.

Separately, a cross-sectional study of phthalate metabolites in the urine indicated positive and significant correlations with abdominal measures of obesity among adults in the USA. Phthalates are ubiquitous in the environment due to extensive use as plasticizers for food contact materials and other industrial applications. Persistence of phthalates adds to environmental and human health concerns over packaging materials and food products containing these compounds.

The foregoing review confirms the general thesis that obesity interacts synergistically to exacerbate lung and cardiovascular impairments in individuals exposed to ambient air pollutants. This is a significant conclusion in view of the obesity epidemic overtaking society and the unrelenting pollution in urban settings where most of the population live.

There is already reliable evidence that obesity induces a state of low-grade inflammation. In addition, high levels of reactive oxygen species (ROS) are inextricably linked to obesity and related pathologies, particularly insulin resistance and development of type 2 diabetes. These pro-oxidants can damage cellular proteins, lipids and nucleic acids, causing functional disorders. It is further postulated that DNA methylation signatures may be associated with adiposity and obesity, thus implying an epigenetic association in individuals with a high BMI.

In addition, it is proposed that alterations in microRNA expression may induce changes in the array of genes regulating a diverse range of biochemical pathways, including adipogenesis, inflammation in fatty tissues, lipid metabolism and insulin resistance.

Another important aspect of obesity is reflected in effects on immunocompetence. Obesity is associated with stress and dysfunction of adipose tissue as well as of the liver, skeletal muscle and pancreas. These processes lead to fat accumulation in primary lymphoid organs such as bone marrow and thymus. As a consequence, there is a loss of lymphoid tissue architecture and integrity, leading to altered distribution of leukocyte subsets and populations. The net result is that immunity is impaired in obese individuals.

A wide range of pollutants also exert effects similar to those existing in obese individuals. Thus, the inflammatory and oxidative stress responses as well as the epigenctic modifications and immunological impacts are repeatedly invoked to characterize the physiological effects of both POPs and ambient air pollutants. It is possible, therefore, that pollutants may upregulate separate components of the same pathways pre-existing in obese individuals to induce synergism in the onset and outcomes associated with disorders such as COPD, asthma and cardiovascular disease.

- What is 'insulin resistance'?
- Write a short essay on the immunotoxic effects of pollutants.

10.3.5. The human health impacts of environmental carcinogens vary according to source and type of stressor, as indicated by Table 10.2.

However, the IARC classification requires amplification in terms of specific and multiple sites of malignancy, as outlined below:

- mycotoxins:
  - aflatoxins:
    - liver cancer;
  - fumonisins:

Table 10.2. The IARC classification of food and environmental carcinogens (selected examples).[a]

Group	Definition	Carcinogen
1	Carcinogenic	Aflatoxins; particulate matter; PCBs; dioxins; lindane; shale oils; arsenic; cadmium; fission products; radon; UV radiation
2A	Probably carcinogenic	Polybrominated biphenyls; dieldrin; malathion; glyphosate; lead (inorganic)
2B	Possibly carcinogenic	Fumonisin B$_1$; heptachlor; parathion; 2,4-dichlorophenoxyacetic acid; benzophenone
3	Not classifiable for carcinogenic potential	Mercury

[a]IARC, International Agency for Research in Cancer; PCBs, polychlorinated biphenyls; UV, ultraviolet.

- oesophageal cancer;
- toxic air pollutants:
  - IARC classification;
  - nitrogen dioxide:
    - lung cancer;
  - particulates;
  - benzene, toluene, ethylbenzene and xylene (BTEX);
  - acetaldehyde;
- ionizing radiation:
  - atomic bomb exposures:
    - breast cancer;
    - liver cancer;
  - Chernobyl accident:
    - thyroid cancer;
- radon:
  - lung cancer;
- ultraviolet (UV) radiation:
  - skin cancer;
- confounding factors:
  - hepatitis virus;
  - human papillomavirus;
  - gender differences; and
- alerts.

In the case of biogenic food contaminants, the aflatoxins and fumonisins have been linked with, respectively, hepatocellular and oseophageal cancer. In addition, there are efforts being undertaken to classify ochratoxin A as a carcinogen on the basis of its formation of adducts with DNA and its induction of oxidative stress and epigenetic modulation.

According to the IARC, urban air pollution, including particulates, are regarded as active carcinogens. There is support for the hypothesis that traffic-generated air pollution may be linked to the development of breast cancer, especially in menopausal women. In addition, an association between ambient nitrogen dioxide and lung cancer incidence has been proposed, with residential proximity to urban and industrial pollution being a significant risk factor.

Volatile organic compounds (VOCs), including BTEX are recognized traffic-associated air pollutants, with benzene representing a significant cancer risk in high-exposure groups. Furthermore, acetaldehyde present in vehicle exhaust emissions can induce DNA adduct formation and may, therefore, contribute to cancer risks associated with urban air pollution. The National Institutes of Health (USA) classify acetaldehyde as 'reasonably anticipated to be a carcinogen'.

Evidence obtained in the aftermath of atomic bomb explosions and nuclear power accidents continues to reinforce the association between radionuclide exposure and cancer. For example, a 2017 analysis indicated that solid cancer risks among atomic bomb survivors in Japan, for the period 1958–2009, remain elevated more than 60 years after exposure.

It is well established that environmental ionizing radiation is a potent risk factor for breast cancer incidence, particularly when exposure occurs at a young age. This knowledge is based largely on observations of the cohort of atomic bomb survivors participating in the Life Span Study in Japan. A 2018 study demonstrated that, more than 60 years after exposure, a strong dose–response relationship for breast cancer was exhibited in both women and men survivors. It was also concluded that, in females, breast tissue sensitivity to radionuclides was increased when exposure occurred during puberty.

A subsequent investigation with this cohort demonstrated significant excess radiation-related risks for the development of liver cancer compared to observations reported in a preceding study. However, it is important to note that chronic infection with the hepatitis virus is also an important risk factor for liver cancer.

In contrast, exposure to radiation in the Chernobyl accident was dominated by the emission of radioiodine, resulting in the induction of thyroid cancer. Inhalation and consumption of foods, particularly milk, contributed harmful intakes of radioiodine. The IARC emphasize that the thyroid gland is one of the organs most vulnerable to cancer induction by radionuclides and the incidence of this type of cancer in Chernobyl survivors is entirely within expectations. However, the evidence indicates that the risks for thyroid cancer are greater for children. For example, in the period 1992–2002, more than 4000 cases were diagnosed among those who were children or adolescents at the time of the accident.

Another contrasting feature in environmental carcinogenesis is seen in the epidemiology of indoor radon exposure. It is widely accepted that residential exposure to radon, at levels above 200 Bq m^{-3}, is a clear risk factor for the incidence of lung cancer, irrespective of confounding factors such as tobacco smoking. However, at radon concentrations below 200 Bq m^{-3}, tobacco smoking and residential proximity to vehicular pollution may complicate conclusions based on epidemiological observations.

The effects of UV radiation provide further evidence of distinctive characteristics in environmental carcinogenesis that are complicated by other factors. UV radiation is essential for vitamin D synthesis in the skin. However, UV exposure can also induce skin cancer which is now regarded as the most prevalent type of malignancy in humans worldwide, particularly affecting elderly male Caucasian individuals. The diary presented in Table 10.1 highlights a recent announcement that skin cancer mortality rates have more than doubled in 50 years especially in young adults. Over 210,000 skin tumours were reported in the UK in 2015. While vitamin D may be protective, as emphasized by Vishlaghi and Lisse (2020), there is evidence that beta human papillomavirus infections together with UV radiation may contribute to the development of cutaneous squamous cell carcinoma, a well-recognized manifestation of skin cancer.

It is clear that evidence of environmental carcinogenesis may continue to emerge several decades after initial exposure, as in the case of nuclear fallout in Japan and Chernobyl. It is also apparent that multiple forms of malignancies occur after exposure to radionuclides in survivors of atomic bomb and accidental explosions, although the effects on thyroid cancer are of particular note in Chernobyl. In contrast UV radiation is associated primarily with skin cancer, albeit with three subtypes. Residential exposure to radon is also linked to a single type of manifestation, specifically in the lungs. It should also be recalled that occupational or residential exposure to metals such as nickel can result in cancers of the respiratory tract.

Finally, the impacts of other factors such as viral infections, vitamin D status and gender may determine initiation and progression of cancer. For example, 2021 experimental data suggest gender differences in UV-induced skin damage.

The question of environmental triggers in carcinogenesis has received enhanced impetus due to the notion that stressors may operate in different ways, in addition to directly causing mutations (see alert in Table 10.1).

- What are the main stages in carcinogenesis?

10.3.6. In 2019 it was estimated that lung cancer incidence, worldwide, results in almost 2 million fatalities each year. It was also estimated that ambient air contaminants annually account for just over 300,000 deaths. There is considerable interest, therefore, in determining the extent to which tobacco smoking and ambient air contaminants may act in the induction and progression of lung malignancies and the likelihood of synergism between these two factors, as set out below:

- It is universally agreed that tobacco smoking is a direct pre-eminent risk factor for lung cancer development.
- A second important risk factor is ambient air pollution in urban environments. For example, particulate matter associated with incomplete combustion of biomass and gasoline is consistently associated with pulmonary neoplasms. Biomass is still widely used for power generation, domestic heating and cooking in Eastern Europe and Indochina.
- Vehicular traffic in cities also contributes significantly to the emission of particulates from combustion of gasoline as well as from tyres and road surfaces. These carbon emissions carry an additional toxic burden in the form of PAHs adhering to the surface of particles.
- Metal toxicity is also implicated in lung malignancies which may be exacerbated by tobacco smoking and environmental contaminants.
- A third risk factor in lung cancer incidence, complicated by tobacco smoking, is exposure to indoor radon.

The model presented in Fig. 10.4 depicts possible environmental–tobacco smoke interactions in the incidence of lung cancer. It can be seen that particle-bound PAHs, VOCs and cadmium are common

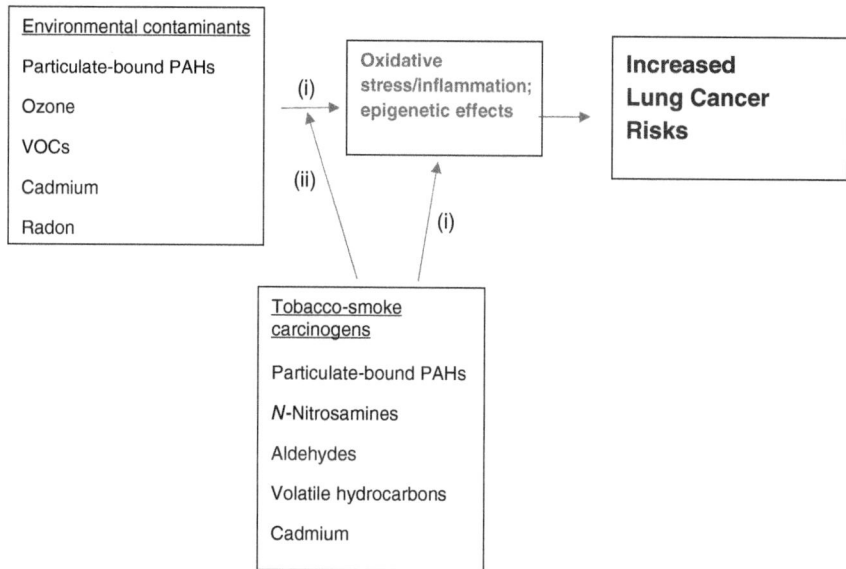

Fig. 10.4. Pathways to lung cancer: the interaction between environmental contaminants and tobacco-smoke toxicants, a hypothetical model; (i) direct effects, (ii) additive or synergistic mechanisms. PAHs, polycyclic aromatic hydrocarbons; VOCs, volatile organic compounds.

- Can you name one VOC?
- What are 'aldehydes'?

carcinogens in atmospheric and tobacco smoke. Although the model is hypothetical rather than definitive, it is based on evidence of individual effects of carcinogenicity and on case studies, as described below.

Both tobacco smoke and solid fuel particulates contain comparable carcinogens and are, therefore, associated with similar health disorders such as COPD and lung cancer. In addition, both types of particulates can induce genome mutations, alternative splicing of mRNAs, dysfunction in epigenomics and initiation of tumour-promoting chronic inflammation. A case study in Xuanwei City (China) provides epidemiological evidence of an association between PM_{10}, $PM_{2.5}$ and lung cancer due to use of smoking coal for domestic heating and cooking in unvented indoor fire-pits. Lung cancer incidence in this city is among the highest in China. It is conceivable that tobacco smoking might contribute to this enhanced incidence of malignancy.

Again, in China, the results of one study indicated high cancer risks from inhalation exposures to PAHs, particularly in winter in the Fenhe Plain. The effects were attributed to traffic exhaust emissions which

215

continue to increase. Another investigation in a new urban district of Nanjing (China) suggested high potential lung cancer risk due to ambient pollution with PAHs. It is possible that smokers in different parts of China, and elsewhere, may be at higher risk of contracting lung cancer due to additional burdens of PAHs interacting with other carcinogens present in air and cigarette smoke (Fig. 10.4).

Metals are important carcinogens in tobacco and cigarette smoke as well as in specific environments. Of these, cadmium is regarded as a potent promoter of a wide range of malignancies, including lung cancer. This element is highly volatile, efficiently transferred to mainstream cigarette smoke and deposited in lungs before transportation to peripheral organs and tissues.

Emerging data indicate that inhaled tobacco-smoke cadmium correlates with a diverse array of pathologies, including lung cancer. Blood, urine and tissue cadmium levels and cadmium:zinc ratios are significantly different between smokers and non-smokers and for several cancers among smokers. Cadmium exposures in humans also occur in severely polluted areas such as the Jinzu River Basin in Toyama (Japan), correlating with mortality due to renal and uterine cancers.

Other investigators maintain that the association between cadmium exposure and lung cancer is well established due to the ability of the lung to retain relatively high quantities of the metal after inhalation. Similar risks may, in addition, arise during recycling of electronic waste, particularly in small-scale units in developing countries. In both of these cases, there may be enhanced risks associated with cumulative exposures to cadmium.

Ambient ozone in urban environments has also been linked to lung cancer incidence and progression. These effects are consistent with the well-established reactions of the gas to oxidize biomolecules, resulting in deleterious tissue inflammation. It is widely recognized that chronic inflammation facilitates lung cancer promotion. Furthermore, it is known that individuals with inflammatory COPD are at significantly higher risk of developing lung cancer than those smokers who do not display this form of pulmonary disease. Work with an animal model indicates that ozone exposure increases metastatic distribution to peripheral organs via enhanced tumour cell migration and attachment to lung tissues. It is, therefore, possible that ambient ozone may act with other urban pollutants as well as with tobacco toxicants in an additive or synergistic mechanism to induce lung cancer.

In a more definitive analysis, Melloni (2020) indicated that radon-linked lung cancer risks in individuals smoking 15–24 cigarettes/day are significantly higher compared to non-smokers. Melloni (2020) concluded that 'combined exposure to cigarette smoke and radon is regarded as synergistic'.

It is thus clear that lung cancer interactions between cigarette smoking and environmental contaminants, while conclusive for radon exposure, require further elucidation. Furthermore, such synergisms may also operate in other forms of cancer, particularly breast cancer.

- What is 'metastatic' distribution?

10.3.7. Case study: 'Pollution and pregnancy outcomes in women'

This news item referred to work at the Mount Sinai Hospital in the USA which focused on the developmental effects of ambient ultrafine particulates in pregnant women living in the metropolitan area of Boston. However, it is important to adopt a more comprehensive approach to the question of prenatal exposures. This should acknowledge a greater diversity of pollutants and developmental outcomes. A summary is provided below:

- ultrafine particles:
  - spontaneous abortion;
  - stillbirths;
  - preterm births;
  - reduced birthweights;
  - lung development;
  - childhood asthma;
  - neurological development;
  - cognition impairments;
  - autism;
  - obesity;
  - diabetes;
- PCBs:
  - the Inuit case;
- dioxins:
  - the Yusho poisoning case;

Continued

10.3.7. Continued.

- pesticides;
- mercury; and
- radioactive fallout:
  - the Chernobyl contamination case.

There is reliable evidence linking ultrafine particulates to adverse developmental outcomes throughout gestation (Johnson *et al.*, 2021). Adverse effects include increased risk of stillbirths, preterm birth and infants with reduced weight at delivery. The third trimester of pregnancy appears to be a critical period for manifestation of these effects. Ultrafine particulate exposure during the third trimester is associated with over 40% increased stillbirth risk. Other evidence implicates prenatal exposure to ambient sulfur dioxide and carbon monoxide in addition to particulates in the incidence of low birthweights in babies.

Prenatal exposure to ultrafine particulates in polluted urban air affects lung development and subsequent respiratory functions that may continue throughout childhood. For example, these pollutants can cause disruption of alveolarization and pulmonary immune development.

In the Boston investigation, 18% of children exposed pre-natally to ultrafine particulates subsequently developed asthma around the age of 3 years, compared to an average incidence of 7% across the USA. This effect was attributed to the entry of ultrafine particulates into regulatory systems to affect neuroendo-crine and immune functions. Such changes often determine subse-quent acute and chronic outcomes in asthmatic children. For example, exacerbations may develop as a result of prenatal exposure to ultrafine particulates and the incidence of pneumonia in these children.

Prenatal exposure to ultrafine particulates is also associated with impairments in behaviour and cognitive skills in children. Additional impacts include the development of metabolic derangements, particu-larly obesity and diabetes.

Concerns emerged in 2020 regarding the exposure of pregnant Inuit women to PCBs and mercury via regular seafood consumption. Despite significant declines due to dietary changes, exposure to these contaminants still occurs among these individuals. Based on 2017 blood plasma data it was estimated that 10% and 20% of pregnant women exceeded national guidance values for PCBs and mercury, respectively.

Continued

10.3.7. Continued.

Exposure to dioxins during pregnancy is also associated with adverse effects in offspring. The Yusho poisoning case in Japan relates to consumption of rice oil contaminated with PCBs, PCDFs and PCDDs. Pregnancy outcomes in affected women included increased miscarriage frequency, premature parturition, reduced birthweights and pigmentation defects in babies. These impacts correlated with elevated maternal blood dioxin status.

A common source of dioxins is emission from municipal and other waste incineration processes. There are consistent concerns over adverse pregnancy outcomes in women living near such installations. This issue remains controversial, often due to lack of statistical power in the data.

Of considerable concern is the incidence of congenital defects associated with maternal exposures to pesticides, particularly atrazine. Risks arise due to occupational use or proximity to arable farms where pesticide contamination often occurs via atmospheric aerosols. A diverse range of abnormalities have been attributed to such exposures, including:

- heart defects;
- hypospadias;
- male genital malformations;
- choanal atresia; and
- stenosis.

Despite the passage of time, several regions in Ukraine, Belarus and Russia remain heavily contaminated with radioactivity in the wake of the Chernobyl disaster. It is regarded by some researchers that this contamination is radiologically significant and likely to persist for decades. Ionizing radiation is widely associated with disruption of embryonic development, expressed in terms of:

- fetal mortality;
- physical abnormalities:
  - limb reduction defects;
- metabolic disorders;
- genetic defects;
- congenital malformation of the central nervous system; and
- Down's syndrome.

Thus, although the 2021 news headline cited in the question is justified by the scientific evidence, it is clear that other developmental end

Continued

10.3.7. Continued.

points, in addition to childhood asthma, deserve equal attention. Furthermore, the impact of prenatal exposures to organic pollutants, mercury and radioactive fallout should be considered, in view of the incidence of a wide range of congenital defects in children. The neuro-developmental impacts of pollutants deserve further attention due to adverse behavioural outcomes in children, as detailed in the answer to question 10.3.8.

- What is a 'trimester'?
- Identify the causal organism of pneumonia.

10.3.8. It is consistently hypothesized that exposures to environmental stressors contribute significantly to incidence of neurological disorders in humans (Chen *et al.*, 2022). The major conditions of relevance here include:

- autism spectrum disorders;
- depression; and
- neurodegenerative disorders.

While genetic predisposition is a primary risk factor for these impairments, a wide range of environmental stressors have also been implicated, including:

- ambient air pollutants:
  - particulates;
  - ozone;
- PCBs;
- pesticides:
  - insecticides;
  - herbicides;
  - fungicides;
  - synergism; and
- noise.

Early-life exposure to air pollution is positively associated with autism spectrum disorders. Investigators in 2020 identified a critical phase between late pregnancy and early postnatal periods for manifestation of adverse effects. Autism spectrum disorders comprise a heterogeneous group of neurological conditions that are characterized by impairments in social interaction and communication as well as incidence of repetitive behaviours.

The exponential increase in incidence of autism has stimulated interest in pollution as an additional risk factor. PM_{10} and $PM_{2.5}$ exposures during the prenatal developmental period have been implicated in the incidence of autism in children, although there is a need to investigate the impacts of other air pollutants. This notion is reinforced by emerging evidence indicating that prenatal air pollution modulates neurodevelopment and behaviour in autism by adversely impacting mitochondrial physiology and functions (Frye *et al.*, 2021).

Previous observations indicated a high prevalence of mitochondrial dysfunction in children with autism. In particular it was suggested that 30–50% of these children presented biomarkers of mitochondrial dysfunction, with up to 80% displaying abnormal electron transport activity in lymphocytes and granulocytes. Other observations indicated that in the vast majority of cases, mitochondrial dysfunction in autistic children is not associated with mutations in mitochondrial genes. It is, therefore, implied that these mitochondrial deficits may be the result of environmental stressors. Frye *et al.* (2021) specifically implicated prenatal $PM_{2.5}$ exposure in the development of mitochondrial dysfunction in autistic children, although this relationship was different for those with and without neurodevelopmental regression.

Exposure to ambient ozone has been linked to neurological impairments, including Alzheimer's disease and Parkinson's disease. In patients with Alzheimer's disease, cognitive deficiencies have been associated with oxidative stress and brain lipid peroxidation, thereby compromising memory and neural networks. In Parkinson's disease, alteration of dopaminergic neurons has also been linked to ambient ozone exposure.

There is increasing evidence that exposure to PCBs during critical neurodevelopmental periods such as early life may be a predisposing risk factor for a variety of neurological derangements. Thus, epidemiological studies have linked early-life exposure to PCBs to autism and attention deficit hyperactivity disorder in children. These observations translate into ongoing concerns over the permanent consequences of neurodevelopmental toxicity linked to POPs in general, particularly pesticides.

Epidemiological data consistently point to an association between pesticide exposure and the incidence of Parkinson's disease (Chapter 4, this volume). In particular, the insecticides chlorpyrifos, dieldrin and rotenone have been linked to this disorder. In the case of herbicides, paraquat, 2,4-dichlorophenoxyacetic acid and trichlorophenoxyacetic acid have long been implicated in Parkinson's disease. Exposures to the fungicides, maneb and ziram are also associated with incidence of this disease, particularly when used in combination with paraquat, indicating a herbicide–fungicide synergism. The latter effects are of particular toxicological concern as both types of pesticides target mechanisms that are specific to plants and fungi.

Chronic noise has long been a cause of irritation to residents living near busy roads and airports. However, 2021 experimental and epidemiological evidence suggest that such exposures may contribute to cognitive impairments and incidence of degenerative dementia. It is suggested that amyloid-β deposition and tau hyperphosphorylation in different brain regions, including the hippocampus and cortex may be signatures of these neurological deficits.

Two alerts emphasize ongoing concerns over pollutant exposures and neurological dysfunction (Table 10.1). In 2020, it was announced that paraquat (and other pesticides) would be exported from the UK to non-EU countries. This herbicide has been banned for use in the EU due to the toxicological issues described above. In 2021, higher air pollution exposure was linked to increased risks for incidence of dementia. This alert was based on findings at the University of Washington (USA) relating to residents living at specific addresses in Seattle.

It is clear, therefore, that despite the lack of mechanistic evidence, there are widespread concerns over the impact of pollutants on neurological dysfunction, particularly in relation to behavioural disorders and Parkinson's disease.

- Can you outline the main features of 'attention deficit hyperactive disorder'?
- Give the short form of '2,4-dichlorophenoxyacetic acid'.

10.3.9. As highlighted in Figs 3.1, 10.2 and 10.4, oxidative stress and inflammation are consistently associated with pollutant-related health disorders in humans.

Oxidative stress is the result of an imbalance between environmental or metabolic pro-oxidants and the innate antioxidant defence system in the body. Key points to note are that:

- Ozone is a powerful oxidant present as an ambient air contaminant in polluted cities.
- Pro-oxidants also arise during cellular metabolism, typically in the forms of ROS and reactive nitrogen species (RNS).
- Nitric oxide is a prominent member of RNS.
- Excess ROS/RNS can react with cellular proteins, lipids and nucleic acids to alter the functional properties of these critical macromolecules.

- These alterations can then facilitate the initiation and progression of deleterious inflammatory reactions.
- This is of relevance in:
    - cardiovascular disease;
    - respiratory disorders; and
    - carcinogenesis.

These alterations are also widely associated with exposures to environmental contaminants. In particular, a study at King's College, London indicated that oxidative stress induced by ambient air pollutants can impact overlapping networks involved with endothelial functions, atherosclerosis and even the autonomic nervous system to affect cardiovascular outcomes. There is considerable evidence now that oxidative stress is central to the induction of adverse cardiac outcomes following exposure to ambient pollutants.

There are also reports that exposure to dioxins and furans is associated with impaired lung function, mediated by oxidative stress. In the case of neurodegenerative disorders such as Alzheimer's disease, it is claimed that oxidative stress is a key factor in the aetiology and pathogenesis of clinical manifestations.

An important feature of chronic diseases affected or exacerbated by ambient pollutants is the induction of inflammation. It is widely recognized that inflammation is a part of a defence response to infection, physical injury and exposure to xenobiotics. It may be viewed as a protective mechanism to localize and eliminate the harmful agent. When endothelial cells, for example, are exposed to PCBs or dioxin-like compounds, inflammatory mechanisms are activated resulting in the production of cytokines such as interleukins, interferon gamma and tumour necrosis factors.

Published evidence indicates that air pollution can cause adverse effects on health via a combination of oxidative stress, inflammation and immune mechanisms. For example, prolonged exposure to ozone generally results in permanent changes to lung architecture via airway remodelling and instigation of oxidative stress, resulting in systemic inflammation and extra-pulmonary lesions.

Patients with underlying illnesses such as asthma and COPD are particularly vulnerable to the adverse effects of ambient air particulates. Diverse mechanisms are implicated in these aspects of particulate toxicology. For example, inflammation of the lungs may release inflammatory mediators (oxidants and cytokines) which enter the circulation thereby promoting inflammatory reactions in the endothelium of blood vessels.

Positive correlations between ambient $PM_{2.5}$ and interleukin-6 production and white blood cell counts have been observed in COPD patients in a 2020 study, with stronger effects in individuals with diabetes, obesity and hypertension. It is now generally accepted that macrophages, neutrophils, eosinophils and T-helper cells exert important roles in inflammatory responses by generating cytokines that act as regulators and effectors. In the case of cardiovascular health, it is postulated that ambient $PM_{2.5}$ may alter expression and profiles of microRNAs and cytokines associated with inflammation. Any microRNA dysregulation may result in the development of a wide range of cardiovascular outcomes.

Inflammation may also predispose to the development of cancer, promoting all stages of tumorigenesis. Carcinogenic hazards have been positively linked to all the major categories of biological compounds, POPs and metallic elements. Oxidative stress and inflammation are processes that are invariably associated with the carcinogenicity of pollutants. These processes are also evident in the toxicity of polychlorinated organic compounds. For example, impaired lung function following exposure to dioxins and furans has been ascribed to the mediating role of oxidative stress mechanisms.

The foregoing evidence indicates that the initial stages in the expression of adverse effects of a wide range of pollutant exposures invariably provokes oxidative stress and inflammation responses. These reactions eventually precipitate manifestations of morbidity in humans. Nevertheless, it is salutary to note the cautionary theme expressed by Mudway *et al.* (2020) who stated that, although oxidative stress and inflammation responses are fundamental to all chronic diseases, the specific impacts of pollutants on these mechanisms need further investigation and refinement.

- Explain the role of metabolomics in establishing inflammation responses in communities affected by pollutants from municipal incinerators and chemical plants.

10.3.10. Once in the body, pollutants can impact a variety of endogenous components that instigate or contribute to metabolic instability, culminating in oxidative stress and inflammation. Common cellular constituents at target sites and organs, exemplifying different levels of structural complexity, include:

- DNA;
- microRNA;
- receptors:
  - aryl hydrocarbon receptor (AhR);
  - muscarinic receptors;
  - steroid hormone receptors;
  - lung receptors;
- mitochondria:
  - effects of particulate matter;
  - exposures to nanoparticles;
- endoplasmic reticulum; and
- enzymes, for example:
  - acetylcholinesterase.

The effects on DNA and microRNA are considered at length elsewhere in this volume. The expression and metabolic roles of enzymes are integrated into the following account of receptors and mitochondria as well as in the preceding chapters of this volume.

Receptors occur in all the major tissues and organs of the body and determine the fate of xenobiotics in humans and other vertebrates. In the case of certain POPs, reaction with the ubiquitous AhR is an important mechanism capable of differentially instigating both detoxification and deleterious pathways. AhR is defined as a ligand-activated transcription factor that responds to endogenous or environmental ligands such as PAHs and other POPs (Vogel *et al.*, 2020). Activation of AhR can result in upregulation of cytochrome P_{450}-metabolizing enzymes which initiate biotransformation of xenobiotics to reduce toxicity or, in the case of PAHs and other compounds, to enhance adverse effects via the production of more toxic metabolites. ROS generated in this process, if uncontrolled, can cause oxidative stress and, therefore, a variety of disease states.

In addition, a 2018 study provided evidence that PAHs present in ambient urban particulates associated with vehicular emissions can drive proinflammatory T-cell and dendritic-cell responses via interactions with AhR. Thus, AhR activity is integrated with immune responses and, in the case of certain POPs, ultimately aggravate inflammatory states.

Other cellular receptors are also implicated in human responses to pollutants, based on research with experimental models. For example, modulation of muscarinic receptors and acetylcholinesterase in brain

225

and peripheral tissues following repeated treatment with organophosphate insecticides has long been recognized. Such exposures cause a decrease in these receptors, suggesting development of tolerance to organophosphate toxicity. These declines may also impact on cognitive dysfunctions associated with chronic exposures to these pesticides.

Effects on steroid hormone receptors mediates the endocrine dysfunction caused by a variety of organic pollutants. For example, hydroxylated metabolites of PCB congeners can act as agonists and/or antagonists of oestrogen receptors. In addition, it has been suggested that these metabolites may act as endocrine disruptors in respect of androgen and glucocorticoid receptors.

The potential role of cadmium and nickel in relation to oestrogen receptor signalling and breast cancer is also under review. The relevance of current interest is due to global increases in the incidence of this malignancy. Both genetic as well as environmental factors have been implicated in the initiation and progression of breast cancer. A hypothesis, yet to be verified, has been advanced that nickel and cadmium may operate by binding to oestrogen receptors, thus simulating the configuration and action of the steroid hormone. Previous correlations have demonstrated that lifetime exposure to endogenous oestrogen in women is a well-established risk factor for the initiation and progression of breast cancer.

Any contaminant that simulates the action of oestrogen is likely to contribute to or promote the incidence of breast cancer, particularly in genetically susceptible women. Nevertheless, it should be noted that metals bind to a wide range of other proteins and cellular components and, thus, the role of nickel and cadmium in breast cancer development remains to be elucidated.

Lung receptors have also been implicated in the cellular response to ozone exposures. For example, ozone-induced lung inflammation and hyperactivity are thought to be mediated via tumour necrosis factor (TNF)-α receptors. The cellular effects of this factor interact with two structurally-related but functionally-disparate receptors, with both being co-expressed on the surface of lung cells.

Mitochondria (Fig. 1.1) are cellular organelles generally associated with energy generation via ATP biosynthesis. Mitochondria are particularly concentrated in ventricular tissue, controlling energy transactions and modulating cell signalling. However, mitochondria also exert important functions in biosynthetic processes, regulation of cell signalling and cell death, generation of ROS in response to environmental stressors as well as production of redox molecules and key metabolites. It is now well established that mitochondria are vital for stress sensing to enable cellular adaptation to the external environment.

It has been consistently demonstrated that these diverse and critical functions also provide focal points for the disruptive actions of common pollutants. For example, particulate matter is a highly significant component of toxic air pollution in urban environments. The adverse effects of particulate matter are often linked to physical criteria, but the adhering chemical contaminants, particularly PAHs and dissolved metals may be more significant contributors to toxicity. Emerging evidence indicates that these constituents of particulate matter determine adverse effects, but the overall outcomes in terms of redox imbalance and regulation are dependent not only on transcription factors but also on mitochondrial metabolism.

Other contemporary data indicate that iron-laden airborne nanoparticles represent plausible and pervasive risk factors for myocardial mitochondrial dysfunction and cardiac oxidative stress. Concerns were highlighted following observations that iron-rich particles within the mitochondrial fraction matched the profiles of particulates in local traffic/industrial sources. Myocardial iron-rich deposition in mitochondria occurred even in young children exposed to urban pollution. Chronic exposures to such nanoparticles, even *in utero*, can subsequently facilitate onset and severity of cardiovascular disease into adulthood. Inhalation of iron-laden particles can result in the synthesis of damaging levels of ROS. High concentrations of bioreactive iron oxides are emitted from brakes and fuel combustion in vehicles. Such exposures may also promote other forms of morbidity in subjects previously compromised by underlying disorders such as cystic fibrosis, asthma and COPD.

Furthermore, neurobiological impairments are consistently associated with air pollution via oxidative stress and mitochondrial dysfunctions. It is proposed that oxidative stress elevations disrupt mitochondrial activity in high bioenergetic sites of the brain, including the hippocampus, amygdala and the prefrontal cortex, causing neuronal and behavioural deficits. In experimental models, mitochondria-targeted antioxidants prevent manifestations of these impairments.

The endoplasmic reticulum is a membranous organelle within cells, driving synthesis and folding of proteins. Additional functions include maintenance of cellular homeostasis and signalling. Consequently, activity of the endoplasmic reticulum is subject to modulation by intrinsic as well as external factors. Endoplasmic reticulum stress is now a recognized feature, associated with diverse conditions such as immune responses, metabolic disorders, pulmonary fibrosis, cancer, cardiovascular disease and neurodegenerative deficits.

There is also interest in ROS–endoplasmic reticulum stress–mitochondrial dysfunction interactions in relation to cardiac arrhythmogenesis. Consistent with the foregoing associations, it is entirely expected that endoplasmic reticulum stress should manifest in response to environmental pollutants. Thus, the results of *in vitro* studies indicate that exposure to particulate matter can induce endoplasmic reticulum stress in trophoblast cells. This may provide at least partial insight into the increased risks associated with the incidence of pre-eclampsia in pregnant women following exposure to airborne particulate matter. Other evidence suggests that PCB metabolites may promote atherosclerosis through lipid accumulation and endoplasmic reticulum stress.

The foregoing account clearly indicates that cellular receptors and particular organelles are impacted by or respond to a number of environmental pollutants. However, the detailed mechanisms that connect these responses to pollutant-impacted conditions such as asthma, COPD and cardiovascular disease remain tentative.

- What is a 'ligand?
- Can you define 'transcription factor'?
- What are 'oestrogen' and 'androgens'?

10.4   Ecological Emergency: Biodiversity in Peril

10.4.1. A number of issues are relevant in this discussion, including:

- adequacy of current risk assessments;
- methodological constraints;
- regional observations;
- habitat changes;
- status in agroecology;
- impact assessment:
  - glyphosate-based herbicides;
  - pesticide resistance;
  - ecological imbalance;
  - case study: Lake Shinji (Japan); and
- light pollution.

It is suggested that reports of an 'insect apocalypse' have been exaggerated in global media coverage. A major contention is that insect ecology in vast swathes of Africa, Asia and South America has yet to be studied in any systematic manner and that anecdotal evidence on its own will not be a sufficient basis for risk assessment. In addition, it is suggested that studies on insect decline have been constrained by biases in sampling and diversity metrics. Given the extreme range and distribution of insect species in different habitats, this line of argument is appealing.

However, Owens *et al.* (2020) referred to steep declines in insect diversity and biomass across geographically diverse regions, including several Western European countries, Puerto Rico and Costa Rica. Whether insect declines are associated with the intensification of arable farming and use of pesticides in these countries will undoubtedly continue to provoke active debate.

It is not readily appreciated that insects are essential components of terrestrial and aquatic food webs thereby underpinning critical ecosystem functions. In addition, insects serve as highly effective pollinators in arable and horticultural enterprises.

Whatever the arguments over the existential narrative, it is patently clear that beneficial insects, for example, are sensitive to habitat changes and pollution in its diverse forms. The risk for these insects is a matter of considerable concern, particularly in managed ecosystems such as arable farms. In 2019, studies at the University of Sussex (UK) indicated that 23 bee and wasp species have been declared extinct over the last century.

The role of pesticides used in crop production is consistently associated with this decline in populations. For example, herbicides can alter the floral diversity available for insect pollinators such as bees, butterflies and hoverflies which may also be affected directly by multiple pesticide residues on these flowers. Recent findings at the University of Wageningen (the Netherlands) highlighted the deleterious effects of the insecticide, fipronil, on survival of honeybees and butterflies.

Other data indicate that sub-lethal concentrations of pesticides may alter the behaviour of social bees and reduce survival of entire colonies. Nevertheless, detractors tend to argue that any negative impacts only occur when applied pesticide concentrations exceed levels found in the pollen and nectar of treated plants. It is also suggested that bees are able to avoid or reduce total pesticide exposure by foraging on other flowers, depending on pesticide drift in treated fields.

229

The continuing debate over glyphosate-based herbicides in relation to carcinogenicity has overshadowed the potential ecological impact of this pesticide. The effects on honeybees encapsulate emerging concerns relating to glyphosate exposure as reviewed by the author during the period 2018–2020. Salient issues include:

- perturbations in the gut microbiota;
- the effect on gut microbiota occurs with the active agent and not its metabolite, amino methyl phosphonic acid;
- sub-lethal doses affect navigation;
- sleep disturbance;
- larval development effects depend on susceptibility of colonies; and
- high toxicity to the larvae of the stingless bee.

Despite continuing disquiet over the impact of pesticides on beneficial species, it will be recalled from the evidence presented in Chapter 4, this volume, that pesticide resistance in target insects is now a major issue associated directly with anthropogenic activity. Consequently, while beneficial species may be in peril in arable ecosystems, undesirable insect pests continue to display diverse mechanisms that circumvent the lethality of pesticides. This ecological imbalance should not be addressed merely by the formulation of more powerful insecticides that might impact adversely on other wildlife species.

A case study based on the adverse impact of the insecticide, imidacloprid, in Lake Shinji (Japan), typifies the continuing risks for arthropods, even with the advent of modern pesticides. Monitoring commenced more than a decade before introduction of the insecticide. However, following agricultural use, imidacloprid seeping into the lake resulted in a collapse in the food web, particularly aquatic insects, causing fishery losses in a commercial enterprise.

Light pollution is emerging as a significant, but underestimated, factor in the decline of insect populations, as emphasized by Owens et al. (2020). It is estimated that over the past three decades, light pollution has doubled in high biodiversity ecosystems.

Artificial light at night affects:

- insect movement;
- foraging;
- reproductive activity: for example, the characteristic bioluminescence signals emitted by fireflies is obliterated by light pollution in cities, meaning that these insects cannot locate each other for mating and propagation of the species;

- predation; and
- development of immature insects.

Whatever the arguments over the 'insect apocalypse', the risks for beneficial insects remains an overriding concern. The impact of pesticides and resistance to these chemicals, the emergence of an ecological imbalance in managed ecosystems as well as the effects of light pollution should receive scrutiny in respect of global populations of insects.

- Define 'microbiota'.
- What is 'predation'?

10.4.2. In common with other aquatic animals, amphibians are at risk due to direct contact with pollutants suspended or dissolved in the surrounding water. Aspects currently under review include:

- skin characteristics;
- oral intake;
- lipophilicity;
- microbiome diversity;
- pesticide impacts; and
- sensitivity to synthetic polymers.

Skin absorption and oral intake are the major routes of entry. According to some sources, amphibians are characterized by a permeable skin that is physiologically involved in osmoregulation and respiration. Uptake of xenobiotics may vary across different skin regions and relative lipophilicity of contaminants may influence percutaneous absorption. Microbiome diversity can differ between skin regions, with pathogenic fungi affecting structure–function relationships in damaged tissues. Consequently, amphibians may be particularly sensitive to environmental contaminants and any adverse effects may be compounded by climate change.

Pesticides in contaminated surface waters present clear risks for amphibians and there is experimental evidence for this notion. For example, acute toxicity determinations with four North American frog species revealed that the surfactant used in glyphosate-based herbicides was more toxic than the different commercial versions of this pesticide. This observation is consistent with evidence for other species presented elsewhere in this volume. It should be noted that there are alerts on the potential detrimental effects of herbicides on the survival and fitness of amphibians in agroecosystems.

Experimental observations also confirm the sensitivity of amphibians to synthetic polymers. For example, following exposure, microplastics have been found in gills, gastrointestinal tract, blood, liver and muscle tissues of tadpoles. In addition, morphological changes were apparent, with reduced ratio between total length and mouth-to-cloaca distance as well as caudal length.

- What is 'osmoregulation'?
- Explain the physiological functions of gills in amphibians.

10.4.3. Bird populations are subject to a wide range of environmental risks due to:

- toxic air constituents;
- pervasive organic compounds in managed ecosystems, including:
  o DDT;
  o Agent Orange;
  o PCBs;
  o PAHs;
  o polybrominated diphenyl ethers (PBDEs);
  o glyphosate;
- crude oil contamination and hydraulic fracturing of shale for fuel extraction:
  o impact of oil spills:
    − Alaska;
    − Vancouver Island;
    − Mauritius;
    − New Zealand;
- radiation;
- anthropogenic sound and light pollution; and
- evidence of risks for:
  o common kestrel;
  o Arctic seabirds;
  o red-crowned cranes;
  o endangered raptors;
  o little egret;
  o warbling vireos;
  o marbled murrelets.

Consequently, impacts are not only diverse but extremely challenging for avian populations in different ecosystems. It is not readily appreciated that air pollution can affect birds as well as humans via inhalation exposures. Consistent adverse impacts have been reported that were attributed to gas-phase and particulate pollutants. Effects noted include respiratory distress, elevated stress levels, behavioural alterations and impaired reproductive outcomes.

In the annals of environmental pollution, the abiding image of harm will, arguably, be exemplified by the extensive spraying of crops with DDT in the 1950s. With the emergence of toxicological data, DDT was prohibited in Europe and the USA, although the pesticide remains in use for the control of insect vectors of malaria and other human diseases. Nevertheless, the risks for wildlife, including birds, persist, for example in terms of important egg characteristics.

In 2020, it was reported that despite the imposition of the DDT ban in 1986, dichlorodiphenyldichloroethylene (DDE) contamination in the Canary Islands (Spain) remains elevated. Residues continue to negatively impact on eggshell thickness in the common kestrel. This observation corroborates classical evidence relating to eggshell thinning in other birds. Furthermore, studies with falcons in south Texas (USA) indicated contamination of eggs with DDE and oxychlordane, correlating with shell thickness values that were lower than in the pre-DDT era.

Declining trends of dioxins, furans and non-*ortho* PCBs in eggs of Canadian Arctic seabirds, including murre and fulmar, have been reported, with the conclusion that current levels are unlikely to affect reproductive performance in these species. However, DDE residues in these samples together with eggshell thickness measurements would have provided valuable supplementary data for a more comprehensive risk assessment.

Other data indicate a disturbing feature of general bioaccumulation of POPs in birds. For example, in red-crowned cranes in Japan, a wide range of halogenated compounds were detected in muscle lipids. In decreasing order of concentration, these included PCBs, DDT and its metabolites, chlordane-related compounds, hexachlorobenzene (HCB), hexachlorocyclohexanes and polybrominated diphenyl esters.

Evidence relating to the little egret in the Poyang Lake Wetland (China) indicates higher bioaccumulation of PAHs in feathers and eggshells compared to organochlorine pesticides, a feature consistent with contamination profiles in water, soil and food samples. The use of down feathers confirms the value of this methodology for monitoring POPs and organophosphate flame retardants in nestlings of the endangered cinereous vulture. Concentrations of PCBs, PBDEs and DDE were higher in down compared to contour feathers, although the latter had higher levels of lindane. It is postulated that down feathers reflect contaminant profiles in the chick due to maternal transfer via the egg.

Additional ecological insight is provided by glyphosate treatment of managed conifer forest plantations in the USA. Glyphosate application increased turnover of bird species in treated sections. In addition, warbling vireos (deciduous specialists) declined in treated areas, suggesting that these species may be particularly sensitive to glyphosate exposure. Moreover, nesting success of certain species was significantly reduced by this herbicide.

Marine birds are at risk due to regular and worldwide crude oil pollution incidents, generally in pristine ecological settings. The physical effects of surface oiling of feathers and ingestion of toxic hydrocarbons represent particular risk factors for these species. In 2016, marine birds were considered to be at potentially higher risk of exposure to persistent oil pollution on the north-east coast of Vancouver Island (Canada). Chronic oil pollution along the south-west Atlantic coast of South America may exacerbate risks for penguins already burdened with organochlorine compounds and PAHs in hepatic and muscle tissues.

The 2020 oil spill in the pristine ecosystems of Mauritius is an important indicator of continuing pollution incidents affecting marine species. Insight into actual harmful impacts on marine bird populations may be gauged from exposures in the *Exxon Valdez* oil spill off the coast of Alaska in 1989. According to Haycox (2020), although a number of species have recovered, the fate of populations of Kittlitz's murrelets still remains unknown. In the case of marbled murrelets, recovery from acute loss has yet to occur. Similarly, pigeon guillemots were considered to be 'not recovered'. It was estimated that 10–15% of the local population of these birds died as a result of severe oiling. However, an increase in nest predation of chicks contributed to further declines in the population of pigeon guillemots.

Rehabilitation of oiled animals is possible, with favourable outcomes. For example, post-release breeding success has been observed in oil-rehabilitated little blue penguins contaminated after the M/V *Rena* accident off the coast of New Zealand. Most of the potential negative impacts were reversed, although hatching success was reduced in the first season after release of rehabilitated penguins.

It is inevitable that hydraulic fracturing of shale in gas and oil extraction should result in habitat fragmentation. Emerging data from the Bakken region of North Dakota (USA) indicates that of 13 breeding bird species investigated, populations of Sprague's pipit declined significantly as habitat fragmentation increased. This relationship was not seen in populations of the remaining 12 species.

Radioactive pollution affects all living organisms, including free-living birds. There is evidence that species richness, abundance and population density of forest breeding birds declined with increasing levels of radiation in the aftermath of the Chernobyl accident. Differential effects were also observed for birds consuming invertebrates living in the highly contaminated soil surface.

As might be predicted, anthropogenic noise pollution is an important factor for bird communities in urban environments. Effects are considered to be associated with reproductive success, affecting distribution of bird species in disturbed environments.

Artificial light at night is a consistent modulator of the behaviour and physiological responses of birds, with differences apparent between urban and rural populations. In particular, captive birds exposed experimentally to light at night tend to show precocious development of the reproductive system and moult earlier than those in dark regimes. Other data, however, indicate that anthropogenic noise but not artificial light affects behaviour in equatorial birds. The reasoning is that in the tropics, there is minimal seasonal variation in diurnal lighting patterns and, therefore, birds in these regions are less dependent on photoperiod as cues for reproductive behaviours such as song.

• Can you describe the ecology of wetlands?

10.4.4. The toxicological concerns are amply justified in view of existential risks for apex predators, including (among many others):

• sharks;
• seals;
• polar bears;
• killer whales; and
• crocodiles.

Toxicological issues arise due to biomagnification of:

• PCBs;
• PBDEs;
• DDT;
• HCB;
• endocrine disruptors;
• mercury; and
• cadmium.

It is consistently maintained that both marine and freshwater species, including mammals as well as reptiles are under severe threat due to climate change as well as chemical exposures. It is also clear that marine and freshwater ecosystems are severely polluted with organic chemicals, toxic metals, sewage and plastics on macro, micro and nano scales. These contaminants, accumulating in aquatic biota, are consumed by higher vertebrates.

Apex predators, therefore, acquire a significant burden of pollutants as a result of biomagnification within the food chain. In this process, accumulation of pollutants occurs in tolerant organisms at successively higher concentrations with each step in the trophic sequence. The marine food web provides the usual system for biomagnification to occur, given the diversity of pelagic fish species in the trophic hierarchy.

Biomagnification is a particular issue with POPs and other contaminants which are resistant to detoxification or direct excretion from the affected animal. PCBs and PBDEs commonly occur in worldwide pelagic food webs and are subject to biomagnification in different species of predators.

As a result, sharks, seals, polar bears and killer whales accumulate a diverse array of potentially harmful pollutants, particularly in the major lipid compartments of these species. Contemporary evidence indicates, for example, that in Australia and elsewhere, POPs occur in relatively high concentrations in coastal waters bordering farming and urban areas. From contaminated sediments and pelagic food sources these pollutants are biomagnified in marine animals such as threatened elasmobranchs.

In a 2019 study, PCBs, DDTs and HCB were detected in all collected species of rays and sharks, with concentrations at levels known to induce adverse sub-lethal effects (see also Boldrocchi et al., 2022). Of considerable concern is that endocrine-disrupting compounds constituted up to 65% of POPs in these elasmobranchs. However, in coastal habitats around Reunion Island (south-west Indian Ocean) lower concentrations of PCBs and organochlorines were observed in tiger and bull sharks compared to other shark species in the southern hemisphere (Chynel et al., 2021). The differences may be attributed to limited urbanization and industrialization in the region, species variations or differences in habitat and trophic ecology.

In a circumarctic review of contaminants in ringed seals, POPs, mercury and cadmium were identified as pollutants of greatest concern for these species. This example of trophic magnification of pollutants is of considerable significance since seals are important prey animals for killer whales and polar bears, providing not only valuable nutrients but also a complex burden of organic and toxic metal contaminants. As a consequence, killer whales and polar bears carry a heavy burden of toxic substances.

Particular concerns have emerged over the possible extinction of killer whales due to constraints imposed by feeding behaviour and environmental xenobiotics. Authors of a 2020 report crystallize the issue in concluding that killer whales preying on seals rather than fish are at enhanced risks due to higher consumption of a diverse range of pollutants. Emerging evidence provides a potent reminder that PCBs in particular present significant and continuing risks for a wide range of marine mammals in the North Atlantic Ocean (Megson *et al.*, 2022). There is evidence that Arctic-invading killer whales from fish-consuming populations in the North Atlantic may access marine mammal prey in Greenland waters. Significantly higher concentrations of PCBs, chlordanes and DDT have been found in the blubber of these whales, compared to those feeding on fish in the North Atlantic. Contaminants in other species include brominated flame retardants, PCBs, mercury and cadmium.

There is evidence that pollutants such as PCBs and PBDEs are continuing to enter the food chain of killer whales, emphasizing the toxic legacy of anthropogenic activity. Blubber concentrations of PCBs can exceed toxicity thresholds for immunosuppression and severe reproductive impairments for marine mammals. Additionally, high levels of DDE, selenium and mercury have been detected in stranded whales, with mercury concentrations exceeding thresholds for hepatic damage in marine mammals.

Potentially toxic xenobiotics in polar bears include PCBs, chlorinated pesticides, perfluoroalkyl residues, dioxins, furans and mercury. The half-life of various organochlorine classes in East Greenland polar bears varies with age and sex of animals, with estimates ranging from 5 to 21 years. Organochlorine contaminants were represented predominantly by oxychlordane and DDE, with variations following seasonal, temporal and spatial trends.

It has long been established that the presence of pollutants in wildlife is, at least in part, determined by geographical location. In the case of polar bears, this generalization is exemplified by data indicating that animals from the Russian Arctic had higher blood levels of chlordane and DDE than those in locations east or west of this region. These observations imply the presence of a significant pollution source of organochlorine pesticides in the Russian Arctic region. Similarly, two sub-populations of polar bears in Alaska have been identified with different levels of chlorinated, brominated and perfluorinated contamination in the liver.

It is widely believed that the presence of such a diverse array of contaminants in tissue compartments can additively or synergistically compromise health outcomes and reproductive performance in polar bears. Of particular significance is the reduction in penile (baculum) bone density and fragility as well as size of testes in male bears caused by environmental changes and pollutants, resulting in mating and fertilization failure. Climate change can only exacerbate the risks for polar bears and other apex predators.

The toxicological risks for semi-aquatic predators are equally severe. For example, recent mass mortality of crocodiles in the renowned Kruger National Park in South Africa has raised questions about specific pollutants transported via rivers as a result of industrial, mining or agricultural activities in the respective catchment areas. The contaminants implicated in this pollution include heavy metals and organochlorine pesticides. It has been reported that mercury, selenium and copper in crocodile eggs and eggshells occurred at levels of concern.

In free-ranging crocodiles in South Africa, relatively high levels of lead have been observed which, although not associated with overt toxicosis, might be detrimental to egg development and hatchling health. In north-east Mexico, high levels of cadmium, chromium and lead above local pollution regulations were found in scutes and eggs of crocodiles.

There are additional risks in regions where pesticides are extensively used in agriculture and disease-vector control. For example, in Belize, organochlorine pesticide residues of the DDT type have been found in 72 of 96 crocodile caudal scutes tested, with methoxychlor occurring in all 72 samples. As might be expected, higher levels of contamination were associated with crocodiles from lagoon compared to river habitats.

Thus, there is extensive and consistent evidence to indicate widespread exposures of apex predators to a diverse array of pollutants. The disturbing aspect is that the prospects for remediation are extremely limited for both legacy and contemporary pollutants in aquatic ecosystems.

- Distinguish between 'pelagic' and 'demersal' fish in marine ecology.

10.5   Risk Assessment and Regulation

10.5.1. The absence of a formal definition of 'human organization' in the context of environmental toxicology permits the adoption of a flexible approach in answering this question. The author's summary is presented in Table 10.3.

Table 10.3. Environmental risk assessment for different levels of human organization (based on selected examples).

Level of organization	Risk assessment[a]
Molecular and biochemical	DNA methylation in response to pollutants widely reported; acetylcholinesterase inhibition after exposure to organophosphate pesticides; biochemical pathways affected by biogenic toxins
Organelle	Mitochondrial dysfunction induced by gaseous pollutants and nanoparticles
Cell (see also Fig.1.1)	Cellular processes disrupted by environmental carcinogens; macrophages, neutrophils, eosinophils and T-helper cells exert important roles in inflammatory reactions by generating cytokines that act as regulators and effectors in response to ambient pollutants
Tissue	Long-term exposure to ozone associated with damage to lung epithelium
Organ	Lung function in asthmatics worsened by toxic air components
System	Cardiovascular disease aggravated by air pollutants; central nervous system disorders initiated or exacerbated by occupational or residential exposure to certain pesticides
Whole organism	Reduced birthweights of offspring born to mothers exposed to a diverse range of stressors
Individuals	Exacerbation of asthma, COPD and cystic fibrosis in individuals exposed to toxic air pollutants
Domestic	Exposure to radon associated with lung cancer; in damp buildings, fungal spores and toxins promote or exacerbate respiratory disease
Communities	Mercury, lead and cadmium pollution affecting communities in Minamata Bay (Japan), Flint (USA) and Corby (UK), respectively
Populations/global scale	Radionuclides released in Chernobyl (Ukraine) and Fukushima (Japan) causing widespread contamination
Ecosystems	Fertilizer pollution associated with incidence of algal toxins in lakes, lagoons and coastal waters

[a]COPD, chronic obstructive pulmonary disease.

10.5.2. Epigenetics is the study of structural modifications to DNA that do not involve changes to its base sequence. These alterations can be passed on to future generations, thereby causing developmental and other defects in recipient offspring.

Epigenetics is regularly invoked to explain the toxicity of a wide range of environmental contaminants. However, it can also serve as a means of risk assessment, as explained below. Nevertheless, it is important to note that genetic modulation remains a key mechanism for the expression of toxic effects of contaminants.

Emerging examples of epigenetics in environmental toxicology and risk assessment include:

- DNA methylation signatures:
  - cadmium exposures;
  - developmental impacts;
  - monitoring at municipal waste incinerators;
  - correlations in atherosclerosis;
  - adaptations in marine animals; and
- microRNA profiling:
  - lung cancer pathogenesis.

One of the key indicators of epigenetic modification is DNA methylation, a feature which has epitomized progress in this field. Investigators have determined DNA methylation profiles in cord and maternal blood from mothers selected for high and low cadmium status. These results underlined the importance of the epigenome in modulating the developmental impact of exposure to environmental cadmium in particular and other stressors in general.

In another study, it was reported that children residing near a municipal waste incinerator showed increased body concentrations of heavy metals. Dominant genetic and epigenetic modifications were observed in these children, as indicated by DNA methylation correlations with blood levels of cadmium, chromium and lead.

Urinary heavy metals and DNA methylation have also been correlated with other human health conditions. For example, Lin *et al.* (2020) investigated the implications of such associations in relation to the development of atherosclerosis.

Epigenetic investigations are now being extended to xenobiotic exposure and toxicity in wildlife, for example in the Mediterranean fin whale. A relationship was found between contaminants in blubber and DNA methylation profiles. Genes involved in cell differentiation and

240

functions were affected, suggesting skin abnormalities. It was suggested that these changes in DNA methylation represented a rapid adaptation mechanism to marine pollution. It was further implied that classes of contaminants could instigate specific DNA methylation signatures which would be useful in biomonitoring in marine and, indeed, other species vulnerable to pollution.

Another important class of regulators of gene expression at both the transcription and translation levels are the microRNAs. These are short, non-protein-coding RNAs of about 22 nucleotides in length, attributed with the capacity to enable living organisms to respond to variations in extrinsic stressors, including toxic pollutants.

The epigenetic basis of lung cancer is associated principally with changes in the profile of microRNAs, thus affecting cellular processes such as cell development and proliferation, differentiation, growth and apoptosis. Experimental evidence indicates that microRNAs are profoundly dysregulated during lung carcinogenesis. Accordingly, it is proposed that microRNA profiles and activity may serve as an early biomonitoring tool to identify risks for vulnerable individuals exposed to environmental toxicants. The identification and characterization of microRNA signatures that reflect specific exposures would be a significant advance for risk assessment and imposition of legally binding anti-pollution measures, for example in urban environments.

In conclusion, epigenetics is likely to form part of a comprehensive biomonitoring system to evaluate toxicological risks associated with environmental contaminants. At the present time, there is a need to link epigenetic signatures with epidemiological evidence associating specific pollutants with harm.

- Distinguish between mRNA and microRNA.

10.5.3. The monitoring and control system currently in place is based on a hierarchal arrangement of related agencies with different levels of responsibility, legal mandate and jurisdiction. Administration of this system is implemented at four major levels as follows:

- local:
  - air pollutants;
  - waste disposal and recycling;
  - congestion charges in high-traffic areas;
- national:
  - UK:
    - Department of Health and Social Care (DHSC);

- – Department for Environment, Food and Rural Affairs (Defra);
  - ○ USA:
    - – Environmental Protection Agency (EPA);
    - – Food and Drug Administration (FDA);
    - – Centers for Disease Control and Prevention (CDC);
- regional:
  - ○ European Union (EU):
    - – European Environment Agency (EEA);
    - – European Food Safety Authority (EFSA);
- international:
  - ○ World Health Organization (WHO):
    - – International Agency for Research in Cancer (IARC);
  - ○ Food and Agriculture Organization of the United Nations (FAO); and
  - ○ International Atomic Energy Agency (IAEA).

Local authorities in counties and cities provide data on common ambient air pollutants, with nitrogen dioxide serving as an indicator of potential toxicity in high-traffic areas. These municipal agencies also provide waste disposal and recycling facilities for paper, cardboard, plastics, glass and metal cans as well as pedestrianization of city centres. Recently, the introduction of congestion charges in parts of central London has been widely acclaimed to serve as a model for adoption in other cities.

National government bodies have also been established over several decades with strategic and advisory functions. In the UK, the DHSC is charged with advising and supporting government ministers in all aspects of human health, presumably including pollutant-related diseases. Defra is the primary body responsible for monitoring and control of pollution in the UK via the Environment Agency.

Key organizations in pollution surveillance and control, respected worldwide, are the US EPA and the FDA. The CDC is an additional organization within the jurisdiction of the Federal Government of the USA which underpins activities of the EPA and FDA.

Important regional agencies emerged with the establishment of the EU. The primary body dedicated to environmental matters is the EEA which undertakes 'to provide sound, independent information on the environment'. The EFSA operates within this regional framework, in the jurisdiction of the EU, but its findings often have an international dimension in determining standards for commodities imported from countries outside this region.

At the international level, surveillance and advisories are delivered via the WHO and the FAO, widely recognized and respected agencies of the United Nations. The IARC is a specialized unit within the WHO which regularly provides updates on risks of environmentally induced malignancies. The IAEA is a global intergovernmental organization for the safe, secure and peaceful uses of nuclear physics and technology.

The effectiveness of monitoring and regulatory systems in minimizing risks of pollutants is at best variable, emphasizing the need for greater consistency in approach and coordination between environmental and food safety organizations. A number of issues exemplify both successes and limitations of current systems, relating to:

- falsification of vehicle exhaust emissions;
- implementation of WHO guidance for ambient gases and particulates in polluted cities;
- approval of glyphosate-based herbicides; and
- control of plastic pollution in oceans.

A notable accomplishment, with worldwide implications, has been the US EPA finding that a number of well-established German-manufactured vehicles had been fitted with 'defeat software' designed to falsify emissions performance. This issue continues to reverberate in compensation awards to purchasers of these automobiles (Table 10.1).

In contrast, there appears to be a lack of international coordination in the control of urban pollution. For example, reference to Table 10.1 will show that ambient nitrogen dioxide levels in parts of London significantly exceeded EU limits in 2020. Other data indicated that ambient levels of particulate matter in the city surpassed WHO safety maxima.

Any action to reduce private vehicle use in London would be extremely unpopular and undermine the re-election prospects for leaders. Concurrently, there is an important task of integration in the activities of human health and environmental agencies. For example, during a high-profile investigation of pollution-related asthma exacerbation in a child in London (Table 10.1), an expert witness severely admonished Defra and the DHSC for failing to coordinate efforts on toxic air in the city. There was also a failure of these agencies to adequately communicate with the mother on the health risks associated with pollution in the vicinity of her Lewisham residence in London.

The approval of glyphosate-based herbicides also exposed significant historical deficiencies in the regulatory process. The evaluation procedure was brought into sharp focus in 2018 at the conclusion of a glyphosate-cancer trial in the USA. The manufacturer was ordered to pay substantial damages to an operative who had reportedly developed non-Hodgkin's lymphoma following regular exposure to a glyphosate formulation. The diary records a report in August 2020 that 52,500 US claimants now allege development of cancer due to exposure to glyphosate-based herbicides (Table 10.1). It is, therefore, critically important to enquire why the US EPA sanctioned use of this herbicide, given its classification in 2015 as 'probably carcinogenic to humans' by the IARC. This advice was revised in 2016 by a joint WHO/FAO committee to the effect that there were no cancer risks associated with glyphosate residues in food. However, this declaration ignores other routes of exposure to this herbicide, particularly inhalation of herbicide-contaminated aerosols, as might be the case for the said claimants.

Concerns are also emerging over the failure to integrate evidence of toxicity to wildlife species in the EPA approval of glyphosate-based formulations. For example, it has been known since 2004 that frogs can succumb to adverse effects following acute exposure to different commercial preparations of this herbicide.

Plastics in waterways and oceans, as exemplified in the 'Great Pacific Garbage Patch', has long epitomized the negative impact of consumerism in modern society. Recent surveys indicate that the Greenland Sea is a significant reservoir of plastic particles, while similar debris jeopardize important habitats for wading birds in estuarine ecosystems of the Thames (UK). All the major rivers in the UK are polluted with plastic debris, with the Mersey containing microbeads, fibres and fragments at concentrations exceeding those in the Great Pacific Garbage Patch. Despite the comprehensive arrangements of monitoring and regulatory agencies, there is, at present, a distinct absence of concerted actions to control plastic pollution in rivers and oceans across the globe.

In summary, there appears to be an urgent requirement for executive authorities in local and central governments to implement recommendations of the national and international agencies cited above. The existing lack of compliance is harming human health and threatened wildlife species and may well lead to legal challenges in the future. For example, individual or class-action proceedings could be pursued under the violation of Article 2 of the Human Rights Act, the right to life.

Equally, it is critically important that regulatory agencies cooperate and integrate activities to provide comprehensive risk assessments of the toxic substances in our environment. This collaboration is essential for public health protection and for arresting the decline in populations of numerous threatened wildlife species.

- How effective are the environmental monitoring protocols in your town/city?
- Are you intending to visit a pollution monitoring station in your town/city?

10.5.4. All regulatory agencies rely on well-established protocols to make informed decisions about emerging organic compounds or new applications of approved products. For pesticides, when risk assessments indicate reduced usage, regulatory agencies should modify measures accordingly to reflect the toxicity data. In extreme cases where there is sound evidence of harmful effects, after considering all relevant risk reduction protocols, the pesticide or other compound in question should be withdrawn or denied formal registration.

The underlying philosophy should involve an estimation of the nature and likelihood of the incidence of harm in humans and wildlife species, particularly those under threat due to the combined effects of climate change, habitat loss and cumulative pollutants. In practice, as demonstrated throughout this volume, human health effects are considered in isolation, whereas there is a proven requirement for a holistic approach.

In establishing risk assessment profiles for regulatory purposes, it is necessary to consider routes of exposure to the chemical under review, including:

- food and water;
- ambient air/aerosols;
- contact absorption via the skin or eyes;
- internal sources, including *in utero* transfer; and
- occupational/residential factors.

A well-established sequence is generally followed in human health risk assessments, prior to imposition of regulatory measures for a particular compound. The US EPA is widely respected and held in high esteem internationally for its rigorous methodologies in regulation and provision of guidance in matters relating to pollutant toxicology.

The EPA protocols for pesticides epitomize its attention to detail, embodied in three steps:

- initial step, based on:
  - experimental data;

- o   assessment of human health risks;
  - o   validity of dose–response evidence;
- intermediate step, involving:
  - o   frequency of exposure;
  - o   timing;
  - o   levels of contact;
  - o   cumulative risks; and
- final step:
  - o   integration of quantitative evidence;
  - o   safety factors;
  - o   penalty clauses;
  - o   enforcement mechanisms.

An initial consideration is the evaluation of the toxic potential of pesticides for humans and the intrinsic or extrinsic factors that might contribute to this property. In this evaluation process, scientists in regulatory agencies examine data derived from experimental models and provided directly by the manufacturer or via independent laboratories. A comprehensive range of potential human health manifestations are taken into account for the different pesticides under consideration. Typically, these may include contact irritation, allergies, endocrine disruption, birth defects, effects on pulmonary and cardiovascular functions, neurological disorders and carcinogenic potential.

A fundamental aspect of risk assessment is a consideration of a dose–response relationship, as explained in Chapter 1, this volume. This determination recognizes that ultimate adverse effects are a function of both quantity of dose and intrinsic properties of the chemical under investigation. The data obtained in dose–response assays with mammalian models are then used to estimate equivalent levels that might apply to human subjects.

In the next step in risk assessment, regulators consider the frequency, timing and levels of contact with the chemical in question. Air, food and water are regular sources of a diverse range of potentially harmful chemical contaminants, with certain individuals and communities at particular risk. For those residing in polluted urban environments, constant exposure to toxic ambient gases and particulates present serious respiratory and cardiovascular risks that often result in premature mortality.

For individuals heavily reliant on fish, mercury might constitute an important health risk. Others consuming plant-based diets are likely to ingest a wide range of fungal metabolites and pesticides which may be present within or on the surface of vegetables, food grains and pulses. There may also be cumulative risks associated with particular compounds that share a common mechanism of toxicity. This feature is becoming extremely difficult to accommodate within existing models of risk assessment. For example, a diverse array of pollutants may act synergistically as endocrine disruptors or as carcinogens. Standard dose–response criteria may, therefore, not apply in cases involving multiple mechanisms, for example exposures to PCBs, dioxins or lead.

In the final step, dose–response data and exposure assessments are integrated to characterize the overall risk of the chemical under investigation. Assumptions and safety factors are integrated into this evaluation as well as the robustness of the underlying database. These assessments enable regulatory agencies to present conclusions regarding the nature and extent of risk associated with exposure to the chemical in question, for example a pesticide or a toxic metal. Safety factors are used to allow for variations between persons and, for pesticides, the US EPA uses an extra tenfold increase to protect infants and children,

PCBs are significant legacy pollutants, with the potential to induce harmful effects in both humans and wildlife species (Chapter 4, this volume). Although no longer commercially produced in developed economies, the presence of PCBs in materials manufactured prior to prohibition means that national EPAs are required to regulate the management, clean-up and waste disposal of these persistent compounds. In addition, management of PCB-containing products and equipment still in use remains an important function of regulatory agencies. The Toxic Substances Control Act of 1976 empowers the US EPA to require reporting, record keeping, testing and eventual disposal of materials that may contain specific chemicals including PCBs.

To prevent violations of the PCB provisions included in this Act, the US EPA has developed a penalty policy graded according the gravity of the infringement, taking into account the nature, extent and circumstances of the contamination. With respect to PCBs, the US EPA also provides guidance relating to sampling, real-estate contamination, site revitalization, planning for waste management after natural disasters, treatment of wastes as well as use of incineration and alternative technologies for waste disposal. Despite the development of these complex protocols to safeguard human exposures, it should be noted that levels and metabolic recycling of PCBs will continue to contribute to the heavy burden of pollutants in marine predators.

Regarding the imposition of penalties for infringement of environmental regulations, there is scope for more stringent measures, primarily to deter repeat offenders. For example, a major UK energy company incurred a fine of £7000 for discharging crude oil into the North Sea (Table 10.1). This penalty is unlikely to result in genuine contrition or efforts to improve environmental performance by a company responsible for the *Deepwater Horizon* explosion in the Gulf of Mexico in 2010. Equally important is the regular crude oil contamination of pristine marine ecosystems around the globe, with meagre responses from international regulatory agencies. Similar comments can be attached to remediation of plastic pollution in rivers and oceans.

Lead exposures continue to attract medical and ecological attention worldwide, while illustrating further the operational mechanisms of international regulatory agencies. The US EPA provides guidance on sources of atmospheric lead and the temporal and spatial variations that contribute to pollution. In addition, it sets and aims to revise standards according to emerging scientific and technical data. The US EPA also monitors implementation of regulations to ensure that predetermined standards are attained in particular areas.

Regarding water, the agency assists communities to protect supplies from lead contamination while conducting relevant research in the light of the Madison Wisconsin case study. In 'high-priority drinking water challenges' the US EPA provides technical support via its Office of Research and Development. Within this undertaking is an oblique reference to elevated lead levels in Flint, Michigan (USA) which required legal action by the local community (Chapter 6, this volume).

Notwithstanding the best efforts of regulatory authorities, lead surveillance remains an ongoing worldwide issue, particularly in the USA where concerns emerged in 2017 over lead exposure linked to non-paint sources, particularly candy imported from Mexico.

Despite the rigours of the risk assessment protocols employed by regulatory agencies, important anomalies have emerged, in part, due to lack of enforcement instruments to implement corrective measures. In addition, discrepancies in the interpretation of toxicity data and overriding economic priorities have hampered application of legally binding regulations. For example, in the UK, recent estimates suggest that between 28,000 and 36,000 premature fatalities can be attributed to air pollution. However, despite legally defined limits, efforts to control toxic gases and particulate matter have yielded meagre benefits for individuals afflicted with respiratory or cardiovascular disorders.

It should be noted that outdoor air pollutants, including particulate matter, are classified as Group 1 carcinogens by the IARC.

The solitary UK response to these profound statistics is the introduction of a congestion charge in central London. The futility of this course of action is amplified by a recent survey indicating that over 75% of polluted streets in the UK are in London (Table 10.1). In other polluted cities such as New Delhi and Mexico City, national and local authorities have also failed to implement the recommendations of the WHO, the IARC and other environmental agencies regarding the human health exacerbations linked to pollutants.

A prominent anomaly appeared in the 2018 judgement in the USA, at the conclusion of a glyphosate-cancer case. The manufacturer was ordered to pay substantial damages to an operative who, reportedly, developed non-Hodgkin's lymphoma following routine applications of a glyphosate-based herbicide. Records in Table 10.1 show that 52,500 additional claimants in the USA now allege development of cancer due to exposure to this herbicide.

Toxicologists are now enquiring why the US EPA overlooked the IARC conclusion in 2015 that glyphosate was 'probably carcinogenic to humans'. The parent organization of the IARC (WHO) issued a revised statement jointly with the FAO indicating that there were no cancer risks associated with food residues of glyphosate. A 2017 review by the EFSA confirmed that food residues of the herbicide were not associated with public health risks. However, these declarations overlook alternative routes of exposure, particularly inhalation and skin absorption. Future risk assessments should also consider emerging evidence that glyphosate-based herbicides are more toxic than the active ingredient, implying an associative effect of the surfactants used in commercial formulations of this pesticide.

In view of the foregoing arguments, it is also imperative that regulatory agencies, particularly the FAO and WHO, consider the broad implications for rural communities regularly exposed to pesticide aerosols from arable farms and other activities. It should be noted that, despite its toxicity, DDT is still the insecticide of choice for the control of malaria vectors in the tropics.

The absence of a holistic approach in the final step described above provides additional concerns over risk assessment protocols. Thus, use of glyphosate-based herbicides continues in 2020, notwithstanding accumulating evidence of toxicity to beneficial insects. For example, there is experimental evidence that commercial formulations of glyphosate can be lethal to bees, while other results indicate that the herbicide can perturb the gut microbiota of honeybees, thereby affecting activities such as pollination in economically important crop plants. Since glyphosate is an analogue of glycine, there is a need to investigate potential toxicity in birds and reptiles which depend on this amino acid for nitrogen excretion via uric acid.

An additional safety issue is associated with the marked divergence in regulatory measures enacted and enforced across the globe. For example, it has recently emerged that the USA 'lags behind' other major agricultural nations, including the EU, China and Brazil, in prohibiting use of harmful pesticides (Donley, 2019). This discrepancy is attributed to lack of application by the US EPA in instigation of 'non-voluntary cancellations' for toxic products. Thus, withdrawal of a toxic pesticide requires consent by the regulated industry.

It is, therefore, clear that regulatory agencies have formulated measures within a legal framework to safeguard human health against a variety of pollutants, albeit with varying degrees of success. However, it is of immense concern that the reach of these authorities is severely limited and unlikely to result in any tangible redress in marine ecosystems. This limitation can only magnify adverse risks for species already under threat due to climate change and habitat loss.

- List specific regulatory measures you would recommend to reduce pollution in marine ecosystems in your country.

10.5.5. There is a wide range of pesticides that are currently approved by the major international regulatory agencies for use in agriculture and horticulture. The choice includes glyphosate-based herbicides, the widely used soil fumigant, metam sodium, currently under restrictions in parts of the EU, and thiabendazole.

Thiabendazole has been selected here due to its regular use in preservation of fruit and vegetables from fungal diseases causing blight, rot and other forms of spoilage. The US EPA has issued an extensive toxicological profile as part of its standard risk assessment procedures using data from a variety of sources, including pesticide manufacturers. The general aim is to assess the human health and environmental impacts of the pesticide.

To implement provisions of the Food Quality Protection Act of 1996, the EPA also takes into account potential sensitivities of infants and children to the pesticide as well as cumulative risks for the public as a whole. There are particular concerns over any interactions between thiabendazole and other compounds with 'common mechanisms of toxicity'.

Toxicological profiles for thiabendazole have been established on the basis of a number of key criteria, including:

- target organs in laboratory models:
  - thyroid;
  - liver;
  - kidney;
  - spleen;
- dermal toxicity;
- immunotoxicity;
- neurotoxicity;
- genetic toxicology;
- exposure assessments;
- cumulative effects;
- safety factors;
- requirement for tolerance values;
- synergism with other pesticides; and
- risks for insect pollinators.

The US EPA confirmed that the thyroid and liver are the primary target organs in thiabendazole toxicity in mammalian models, including dose-related increases in absolute as well as relative liver weights. Histopathological changes in kidneys and spleen were also observed in a rodent trial. Separate evidence indicated gross and histological changes associated with urinary tract toxicity following dietary exposure of mice to thiabendazole. These results implied endocrine disruption and changes in immune functions.

Emerging data also point to developmental effects associated with combinations of thiabendazole with another fungicide, imazalil. Maternal exposures induced post-weaning neuro-behavioural abnormalities in F_1 progeny, including effects on exploratory and spontaneous responses.

In addition, the incidence of thyroid tumours, exclusively in males, highlighted a potential issue with malignancy. In a formal classification, the US EPA classified thiabendazole as a likely carcinogen at doses high enough to cause disturbance of the thyroid and hormone balance. In practice, therefore, it is assumed that the carcinogenic effect can be discounted for regulatory purposes.

Despite the available evidence in 2016, the US EPA concluded that the use of currently registered products containing thiabendazole in accordance with approved labelling should not pose unreasonable risks of toxicity in humans or the environment. The EPA further stated that the ecological impact of approved formulations of the fungicide was below its level of concern. Consequently, any environmental mitigation was deemed to be unnecessary.

Since lethality is a design feature of pesticides, the inescapable conclusion is that safety, even of approved products such as thiabendazole, will always be conditional and never absolute. This analysis is reinforced by the US EPA declaration that thiabendazole is highly toxic to freshwater and estuarine invertebrates and fish.

In regulatory and toxicological scrutiny of the impact of pesticides it is critical that potential synergistic effects are considered when combinations are used. For example, neonicotinoid insecticides may be used in conjunction with azole fungicides, including thiabendazole, for greater protection of crops. Synergism between these pesticides may impact on the decline in diversity and spatiotemporal distribution of beneficial and economically important insect species. These concerns led Heneberg et al. (2020) to conclude that the adverse effects of neonicotinoid insecticides on pupation and metamorphosis in a crabronid wasp can be further potentiated in combination with other agrochemicals, including fungicides like thiabendazole.

Although the US EPA assumed that thiabendazole does not share a common mechanism of toxicity with other compounds, there may, nevertheless, be instances of synergism operating by alternative pathways. It is instructive to recall evidence of synergism between paraquat, a herbicide, and the fungicide maneb in animal models of neurodegenerative disorders despite the existence of disparate mechanisms (Chapter 4, this volume).

In ecological settings, there are concerns of multiple pesticide residues in pollen, creating risks for foraging honeybees and other insect pollinators. In a 2019 study, 79 different pesticides and metabolites were identified in pollen. Although insecticides and herbicides predominate, thiabendazole is consistently cited as a contaminant, prompting suggestions of a high frequency of potential synergies in pollen.

It is clear that even for registered pesticides such as thiabendazole, there are good grounds for regular scrutiny of emerging evidence on toxicity. In particular, greater reliance should be placed on independent data sets. Recent claims of undue pressure by pesticide manufacturers have undermined confidence in the impartiality of regulators.

The question of self-disclosure failings should also be addressed to restore integrity in the pesticide evaluation process.

- Why are soil fumigants used by farmers?

10.6 Policy

10.6.1. The 'levelling up' political pledge, expounded in the UK, was aimed at addressing social and economic disparities between the affluent South East of England and the disadvantaged North of the country. The delivery of this policy partly depends on the construction of a high-speed rail link between London and provincial business hubs. However, initial implementation of this project has already resulted in the destruction of ancient woodland, contributing to loss of habitats for wildlife along the route. Thus, provision of economic benefits for the North of England will create ecological and aesthetic disparities that are unlikely to be addressed as part of this programme.

At the global level there is a critical need to address human health and ecological inequalities imposed by environmental pollution. The need for levelling up can be exemplified in relation to different classes of contaminants including:

- harmful ambient gases and particulate matter, with impacts on:
  - childhood asthma;
  - COPD;
  - cardiovascular disease;
- activities at petrochemical plants:
  - gas flaring:
    - vibration;
    - odour;
    - noise;
    - sleep deprivation;
  - 'cancer alley', Louisiana (USA):
    - malignancy;
  - indigenous communities:
    - multiple illnesses;
- toxic metals:
  - lead poisoning;
  - electronic waste toxicity; and
- plastic debris, for example in the Pacific Basin:
  - severe risks for marine species.

Social and ethnic inequalities in exposure to vehicular and industrial air pollution remain formidable issues in polluted cities such as London, Mexico City, New York City, Los Angeles, New Delhi and many other conurbations around the world. Residents in central London and other metropolitan areas are often exposed to levels of ambient air pollutants that exceed legal limits and exacerbate chronic and life-threatening illnesses, particularly childhood asthma, COPD and cardiovascular disease. The case for aligning environmental objectives with those prescribed in WHO guidelines is stronger now than in the past, as epidemiological evidence unequivocally confirms the correlation between toxic ambient air pollutants and a variety of disease conditions. Premature deaths are more likely in these polluted areas than in the verdant suburbs of the same city. Ethnicity is often cited as a major factor in the incidence of morbidity and mortality in polluted cities.

A Europe-wide analysis published by Richardson *et al.* (2013) highlighted specific effects of ambient particulate matter pollution on human health inequalities. It was observed that levels of particulates were strongly related to mortality outcomes in Eastern compared to Western Europe. It is likely that this difference is due to higher levels of pollution caused by the use of coal for power generation and domestic heating in countries such as Poland.

Environmental health disparities are prevalent in other locations. Activities at gas and petrochemical plants often impact on the health and well-being of nearby residents due to diverse sources of pollution. For example, during 2019, residents near to gas flaring operations at a petrochemical plant in Fife (Scotland) complained about vibration, noise, light, odour and smoke pollution leading to health manifestations such as sleep deprivation, headaches, sore throat and asthma.

In Louisiana (USA) a region has been nominated as 'cancer alley', due to 50% higher malignancy rates attributed to chloroprene and ethylene oxide emissions from petrochemical plants in the vicinity of residential estates. Chloroprene is classified by the EPA in the USA as a 'likely human carcinogen', but the issue is being contested by the petrochemical companies and remains unresolved.

In 2021, contamination of air, aquifers, rivers and soil, associated with multiple cases of asthma, cardiovascular disease and industry-specific cancers were reported for indigenous communities in Oklahoma (USA), attributed to nearby oil and gas activities by national corporations. This claim reflects worldwide environmental inequalities and deprivation for native communities in Australia and Brazil in connection with mining pollution.

Ongoing regional disparities for toxic exposures have also been reported in cases of metal pollution, causing unacceptable risks for affected communities. For example, in 2017, it was reported that lead poisoning was higher in parts of California than in Flint, Michigan, with the USA now providing important case studies in modern lead intoxication. Metal pollution associated with electronic waste recycling remains a serious and expanding problem, particularly in developing countries. The range of elements emitted includes lead, mercury, cadmium, chromium, nickel, copper, zinc, aluminium and cobalt. These metals escape into the environment due to informal and unorthodox technologies employed in the manual dismantling, open combustion for metal recovery and illegal deposition of discarded fractions at sites exposed to inclement weather conditions. At abandoned sites, surface soil contamination can exceed regional safety guideline levels. These activities are conducted primarily in developing Asian countries, in consignments exported from affluent nations.

Plastic pollution is another issue requiring levelling-up measures at the international level for the benefit of freshwater and marine wildlife species. Despite the visual nature of this debris and high levels of publicity, particularly in respect of the Great Pacific Garbage Patch, the prospects for remediation in the near term appear remote, given the inexorable rise in consumerism across the globe.

Resolving these pollution-based inequalities is almost certainly not on the immediate agenda of the levelling-up policy expounded in the UK. As at 2021, there appears to be meagre progress in governmental or international efforts to reduce pollution differentials and improve outcomes for the disadvantaged in society or for threatened wildlife species. It is likely that translating the idealism of levelling up into realism might simply lead to promulgation of more platitudes. It will then be the responsibility of affected individuals and communities to pursue environmental justice through the courts, as in the case of lead poisoning in Flint, Michigan (USA) and cadmium contamination in Corby, Northamptonshire (UK). Legal procedures have recently been instigated by residents to reverse the deteriorating ecology and amenities in Lake Erie (Canada).

- How can international interventions help in the levelling-up philosophy?

- Is the proposed siting of new nuclear power plants in Europe and North America likely to lead to environmental health disparities?

10.6.2. This suggestion is based principally on the urgent and entirely valid need to curb global methane emissions from intensive livestock production. In combination with carbon dioxide, methane is implicated as a major contributor to global warming.

However, adoption of this policy would be constrained by profound resource and environmental issues and there might also be adverse nutritional impacts in humans. These issues deserve detailed consideration as follows:

- resource implications:
  - land:
    - increased deforestation;
    - smoke pollution;
  - water;
- environmental risks:
  - increased use of:
    - fertilizers;
    - pesticides;
  - adverse effects:
    - human morbidity;
    - biodiversity;
    - algal blooms; and
- nutritional impacts:
  - mineral deficiencies;
  - excess consumption of high-starch foods:
    - obesity risks;
  - importance of balanced diets.

Increased reliance on plant-based foods would result in greater need for arable land and natural grasslands might be ploughed up for this purpose. However, political leaders and farmers in Brazil, Indonesia and elsewhere would be emboldened to continue extensive deforestation for increased cultivation of soybeans and other staples. Consequently, this policy would adversely impact attempts to achieve global climate change goals. Furthermore, the practice of setting forests on fire to clear land for agricultural use has been associated with unacceptable levels of smog in nearby conurbations. Particulates are now firmly linked to poor health outcomes in affected communities.

It should also be noted that increased crop production would place a significant demand for irrigation in regions where water resources are severely restricted due to drought. In addition, changes in arable farming, including certain conservation measures, have inadvertently resulted in the development of harmful algal blooms in different regions of the world including, for example, Lake Erie (Canada) and the Mediterranean Basin. This effect is attributed to increased agricultural phosphorus and nitrogen loading from tributaries draining into water bodies. Farm pollution is also responsible to a large extent for the destruction of coral reefs around the world.

Moreover, increased crop production would inevitably result in greater use of pesticides, particularly if monoculture is the main component of this strategy. The loss of beneficial insects and other species such as amphibians is a matter of considerable concern, particularly in managed ecosystems. The role of pesticides used in crop production is consistently associated with this decline in populations. For example, herbicides can alter the floral diversity available for insect pollinators such as bees, butterflies and hoverflies which may also be affected directly by multiple pesticide residues on these flowers. Recent findings at the University of Wageningen (the Netherlands) highlighted the deleterious effects of the insecticide, fipronil, on survival of honeybees and butterflies.

There are also alerts on the potential detrimental effects of herbicides on the fitness and survival of anurans in agroecosystems. In 2019, work at the University of Saskatchewan implicated a common farm pesticide, imidacloprid, in weight loss and delayed migration in white-crowned sparrows. Of considerable concern is the development of 'herbicide-ready' concept, which, in effect is a euphemism for transgenic manipulation. Such an idea would enable the development of crop varieties designed to withstand ever higher levels of a particular herbicide, while many weeds develop resistance to the same agents.

Finally, increased reliance on plant-based foods might be associated with nutritional deficiencies of calcium and phosphorus and such diets would require supplementation with animal products such as milk. This fortification would be particularly important for infants and young children. Furthermore, excess consumption of starchy and sugary foods in the forms of pizzas, potato fries, doughnuts, cakes and chocolate are indisputably associated with increased incidence of obesity, particularly in children.

In conclusion, great care must be taken when advocating any change to plant-based dietary regimes.

- Explain the role of calcium in human nutrition.

10.7 Compliance

10.7.1. It is self-evident that compliance with regulations and laws is essential in order to protect human health, fragile habitats and wildlife species. Systems are in place, particularly in the USA and EU to monitor compliance and enforce remedial measures accordingly. Compliance and enforcement are the tools that environmental authorities use to ensure that public services, businesses and industry understand and adhere to pollution laws and regulations.

The sequence in the US EPA compliance and enforcement schedule includes:

- identification of the environmental problem;
- enactment of a law to address the environmental issue;
- implementation of the law via specific US EPA regulations;
- assistance to advise the relevant sectors to understand and comply with US EPA regulations. This process may involve provision of resources and training. The EPA encourages the regulated sectors to assess its obligations and to self-disclose non-compliance, while emphasizing that voluntary admissions with corrective measures can reduce or eliminate certain penalties;
- compliance monitoring through inspections and other mechanisms. On-site visits by qualified inspectors is one option available to the US EPA (and international agencies); and
- enforcement of regulations in instances of non-compliance or when clean-up is required.

In view of continuing exposures, it is appropriate to consider the compliance obligations relating to lead. Separate US EPA directives have been published to emphasize wide-ranging issues that require compliance. For example, for paint, dust and soil, the specific issues relate to new Federal lead-based paint requirements, disclosure of information about such paints, with specific instructions for the construction industry.

Regarding the water industry, explanation is provided for lead and copper rules, laboratory and monitoring procedures and the national pollutant discharge elimination system. For atmospheric lead, references include national emission standards for hazardous air contaminants, the state implementation plan toolkit and air mobile sources programme. Regarding lead waste, relevant resources for compliance include the Superfund remediation system, brownfields and land reclamation, monitoring for hazardous waste units, importing and exporting hazardous waste, managing environmental responsibilities and the construction industry assistance centre.

Despite the aforementioned regulations, which have been in force for a number of years, Katner and Mielke (2020) maintained that lead poisoning is a 'lingering public health priority' as evidence accumulates to demonstrate adverse low-dose risks. The concept based on the statement that 'no level of lead appears to be safe' for children was emphasized previously by the CDC in the USA.

In the global context, there are persistent concerns over crude oil pollution in fragile marine ecosystems (Table 10.1; Pavlov *et al*., 2022). Given the long history of ecological disasters dating back to the *Torrey Canyon* oil spill of 1967, it is believed that industrial and environmental regulations should have been strengthened and explained to the oil and gas industries. Consistent with these failures is the finding that the *Deepwater Horizon* accident was predicated by a series of non-compliance issues relating, for example, to absence of formal risk assessments of critical operations, contingency and mitigation measures prior to the event.

Of particular significance is the aforementioned clause that regulated sectors should self-disclose cases of non-compliance. As shown in the diary (Table 10.1), the effectiveness of this strategy is unpredictable. Thus, the radiation spike at a nuclear reactor in Finland was admitted to the IAEA, consistent with expectations in a highly regulated industry. In contrast, the North Europe radiation spike linked to an installation in Russia remains shrouded in mystery and denials. It is not clear how compliance can be verified or enforced by the IAEA. Similarly, two major US-based international companies appear to be 'dodging environmental responsibility' for electronic waste disposal and presumably operating without complying with safety regulations.

Non-compliance with environmental regulations and industrial safety protocols should, in cases of major pollution incidents, incur severe penalties, as implied in the US EPA directives summarized above. The EPA has reviewed the extent of punitive measures levelled against the companies associated with the *Deepwater Horizon* accident. The unprecedented financial penalties imposed on these companies are now distributed under US EPA supervision and consolidated within the Gulf Coast Ecosystem Restoration Council, created as an independent Federal agency. This council is charged with administering the civil settlements associated with the *Deepwater Horizon* oil spill.

It is clear, therefore, that a sophisticated and comprehensive system of measures is employed by the US EPA to ensure that the relevant sectors understand and comply with regulations and directives. Similar arrangements of protocols are implemented by other national and regional environmental agencies such as the EEA.

Nevertheless, it is abundantly clear that compliance with regulations cannot be assessed or enforced in the marine ecosystem, and consequently pollutants in the biomagnification chain will continue to impact apex predators for the foreseeable future. It is also clear that self-disclosure is only partially effective and that compliance is best achieved by proactive management rather than retrospectively, as in the cases of the *Exxon Valdez*, *Deepwater Horizon* and Mauritius oil spills (Table 10.1). There are, therefore, overwhelming reasons for continued vigilance as regards adherence to regulations in all sectors of industrial activity.

- How was the Superfund used in the *Deepwater Horizon* disaster in the Gulf of Mexico?
- Have you discussed the Mauritius oil disaster in your tutorials? What are the ecological implications? Are there any economic impacts?

10.7.2. Vigilance is a key prerequisite for successful implementation of environmental protection programmes, requiring predictive acumen in conjunction with timely detection. This strategy should provide prompt alerts of impending risks but is dependent on a network of sentinels distributed in communities likely to be impacted by new or emerging activities.

There is a variety of reasons for enhanced monitoring of business activities of farmers, small and medium-size enterprises and multinational conglomerates due to:

- pressures for increased profitability and disbursements of enhanced dividends to shareholders;
- persistent offenders;
- incomplete risk assessments and emerging developments in environmental toxicology;
- inappropriate industrial pressure in securing approval of actual or potentially harmful products;

- climate change factors resulting in excessive precipitation events, waterlogging and run-off of fertilizers and pesticides in effluents from farmland;
- flooding episodes following excessive precipitation causing sewage contamination of rivers and coastal waters;
- failure of self-reporting requirements as advocated/instructed by environmental protection agencies, requiring the intervention of the media and whistle-blowers for identification and resolution of toxicological issues; and
- technological innovations and sentinel functions. These developments have significantly enhanced opportunities for timely detection of pollution events and impending health and ecological risks.

The US EPA regulatory scheme is the global benchmark in environmental protection, serving as an important reference point. Elements of this scheme have been enforced in high-profile cases of misconduct attributed, at least in part, to effective vigilance and widespread awareness of double standards practised by utility companies, multinational corporations and even government departments.

Sentinels based in the popular media have signalled timely alerts in environmental toxicity. For example, in 2021 press reports indicated that the UK government via Defra had sanctioned the application of a bee-killing neonicotinoid insecticide on sugarbeet within a 'limited and controlled' context (Table 10.1).

Significantly, pesticides are often lethal to beneficial insects, amphibians and birds by diverse and complex biochemical as well as physiological mechanisms. For example, a 2013 investigation led authors to conclude that immune suppression by neonicotinoid insecticides may be at the 'root of global wildlife declines'. While this statement is undoubtedly a sweeping generalization, ignoring the impacts of other physiological mechanisms, pesticides and climate change, there are nevertheless grounds for continuing vigilance and concerns over anthropogenic chemicals. Similarly, press reports highlighted the proposed export from the UK of toxic pesticides that are prohibited for use in the EU. The pesticides include paraquat and 1,3-dichloropropene, implicated in neurodegenerative disorders and cancer, respectively (Table 10.1).

Vigilance is also required to expose important and often persistent deficiencies in the quality control measures of national utility companies (Table 10.1). For example, a recent survey concluded that all rivers and lakes in England are polluted with sewage, fertilizers, pesticides, slurry and industrial chemicals. It is clear that despite severe penalties imposed on a major UK utility company discharging untreated sewage into the River Thames in a case that was labelled an 'environmental disaster', pollution of waterways is still a major issue.

New technologies are emerging as sentinels, facilitating the process of vigilance in diverse aspects of pollution incidents (Laneve et al., 2022). For example, satellite images have highlighted risks due to particulates and PAHs for residents within 20 km of a wildfire in the Arctic (Table 10.1). Other images show the Great Barrier Reef over two successive years (2016 and 2017) under threat due to bleaching caused by climate change, pollution and bacterial diseases. The appearance of a 'radiation cloud' presumed to originate from a nuclear fuel plant in the Urals was widely publicized but remains unexplained, emphasizing the unreliability of self-reporting procedures allowed by regulatory agencies. Similarly, the reluctance of authorities in Belgium to delay publication of results of a probe into pesticide-contaminated eggs has further undermined public confidence in self-disclosure strategies as a whole.

The value of vigilance is restricted unless followed by prompt and proportionate actions. A striking example emerged in an investigation conducted by the US EPA which concluded that a number of well-known German-manufactured automobiles had been fitted with 'defeat software'. This innovation was designed to falsify performance relating to nitrogen dioxide emissions. The punitive measures imposed on the offending companies were compounded by compulsory compensation payments to owners of these vehicles (Table 10.1).

In conclusion, self-disclosure of pollution incidents is of limited value in environmental protection and management. Vigilance through the use of appropriate sentinels based in communities as well as use of remote sensing offer significant potential for timely detection of pollution risks.

- Vigilance requires predictive capacity and this may be relatively easy for persistent offenders. Can you identify particular industries that regularly pollute rivers and marine ecosystems in your country?

10.8   Conclusions

10.8.1. There is indisputable evidence that the diverse range of pollutants occurring in ambient air, soil, water and the trophic chain adversely impacts human health and imposes existential risks for wildlife species already threatened by climate change and habitat degradation.

Salient issues relate to:

- Ambient pollutants, including nitrogen dioxide, sulfur dioxide, ozone and particulates, are unequivocally associated with the initiation or exacerbation of respiratory and cardiovascular diseases. The risks are magnified for individuals and communities in toxic urban environments. Regulations for ambient nitrogen dioxide and particulate matter are in force but rarely applied when violations occur. The onus is on pollution-affected individuals to seek redress.
- Genetic predisposition is a contributory risk factor, particularly for the incidence of breast cancer in women exposed to particulates and toxic metals. The role of ethnicity requires elucidation.
- Obesity is another important determinant of ill-health caused by toxic ambient air pollutants. This notion is consistent with outcomes associated with COVID-19 infections.
- Tobacco smoking may compound adverse effects for individuals exposed to toxic air pollutants and radon.
- The causal relationship between UV exposure and the incidence of skin cancer is unmistakable.
- The emissions of radionuclides following deliberate actions or accidental events have delivered a continuing and toxic transgenerational legacy for impacted communities.
- Although nuclear power plants offer the prospect of environmental decarbonization, low-level and accidental emissions of ionizing radiation as well as storage of highly toxic waste present long-term human health risks.
- The toxicity of pesticides has been severely underestimated in terms of unintended consequences for human health and existential risks for wildlife.
- Insect, bird and amphibian populations are at profound risks due to the combined effects of habitat changes and the use of toxic pesticides.
- The notion of a 'safe' pesticide is an illusion that has to be dispelled since it is based on an incomplete assessment of the scientific evidence. The proposed export of toxic EU-prohibited pesticides from the UK to other countries is scientifically and ethically indefensible.
- Marine and freshwater predators such as killer whales, polar bears and crocodiles carry a heavy burden of multiple toxins, including

DDE, lead and mercury. These residues may compound the existential risks imposed by climate change and habitat loss.

10.8.2. A comprehensive spectrum of recommendations is envisaged
to encompass continuing and widespread contamination of the environment with anthropogenic sources of proven and potentially toxic
pollutants.

General

- The primary aim of any remediation process must commence
  with reducing demand at every stage of the pollution process, as
  detailed below.
- Curbing human population growth is a fundamental and obvious
  step in this direction.
- A pollution tax on vehicles, fuel, road use, electronic hardware and
  plastics would be an effective strategy to control emissions of toxic
  gases, metals and particulate matter in urban environments as well
  as reducing pollution in major waterways and oceans.
- Regarding toxic air pollution, affected communities, individuals
  with underlying respiratory and cardiovascular diseases as well as
  local authorities should be aware of the 'duty of care' obligations in
  the event that legally binding limits for ambient gases and particulates
  are exceeded.
- Assessment of ambient air pollution should include ozone and
  sulfur dioxide in addition to nitrogen oxides and particulate matter.
- It is essential that EU/UK/US EPA directives for ambient air pollutants are enshrined in law.
- The rehousing of city-centre residents with underlying respiratory
  and cardiovascular diseases to less polluted environments is a critical requirement, particularly for children, the elderly and cigarette
  smokers.

Specific

- The obesity epidemic across the globe should be urgently
  addressed, as increased BMI correlates consistently with adverse
  reactions to common pollutants.
- Critical evaluation and reduced use of pesticides, other agrochemicals and industrial metals should be considered in view of confirmed adverse implications for human health and wildlife.

- The diary (Table 10.1) shows that efforts should be increased to monitor and enforce compliance in the petroleum and nuclear power industries. Development of contingency measures for environmental emergencies should be critically reviewed and on a regular basis (Callen-Kovtunova and Homma, 2022).
- Prior to commissioning of new nuclear power plants, it is critically important that stakeholders and particularly local residents are aware of unresolved issues regarding decommissioning, clean-out and long-term storage of highly hazardous radioactive waste.
- Concerted action is required to evaluate the toxic impacts of POPs and the role of endocrine-disrupting contaminants on human health and alarming declines in wildlife populations.
- There is, however, an urgent need to explore epigenetic and other means to identify signature molecules and non-random mutations (Grey Monroe *et al.*, 2022) to resolve cause-and-effect issues for pollutant impacts on human health and biodiversity.
- It is essential that the individual agencies for pesticide approval, food safety and environmental protection act together rather than separately and without compromising the well-being of humans and wildlife.
- Risk assessments and regulations are essential to quantify the relationship between specific pollutants and increasing incidence and/or exacerbation of neurodegenerative disorders.
- Education of schoolchildren on issues concerning transport pollution and recycling of consumables, including plastics and electronic devices should be considered.

Redress

- Ultimately, when persuasion and all reasonable actions fail, individuals and communities should seek redress in decisions that can be enforced by the judiciary. There is ample precedence of the success of the legal process, for example in the lead poisoning incident in Michigan (USA), the landmark ruling over the premature death of an asthmatic child in London (UK) and the *Deepwater Horizon* accident (Gulf of Mexico). These radical decisions should embolden individuals adversely affected by pollutants to pursue class actions under local and international directives, particularly Article 2 of the Human Rights Act, the right to life.

- 'Biodiversity Rights Charter': International statutory instruments should be invoked or instituted to protect fragile habitats and wild-life species threatened by persistent anthropogenic toxicants.

10.9 References

Boldrocchi, G., Spanu, D., Polesello, S., Valsecchi, S., Garibaldi, F. *et al.* (2022) Legacy and emerging contaminants in the endangered filter feeder basking shark *Cetorhinus maximus. Marine Pollution Bulletin* 176: 113466. https://doi.org/10.1016/j.marpolbul.2022.113466

Callen-Kovtunova, J. and Homma, T. (2022) Ten years since the Fukushima Daiichi NPP disaster: what is important when protecting the population from a multi-faceted technological disaster. *International Journal of Disaster Risk Reduction* 70: 102746. https://doi.org/10.1016/j.ijdrr.2021.102746

Caverly, L.J., Hoffman, L.R. and Zemanick, E.T. (2022) Microbiome in cystic fibrosis. In: Huang, Y.J. (ed.) *The Microbiome in Respiratory Disease.* Springer Nature, Cham, Switzerland, pp. 147–177. https://doi.org/10.1007/978-3-030-87104-8_6

Chen, H., Wang, K., Scheperjans, F. and Killinger, B. (2022) Environmental triggers of Parkinson's disease – implications of the Braak and dual-hit hypotheses. *Neurobiology of Disease* 163: 105601. https://doi.org/10.1016/j.nbd.2021.105601

Chynel, M., Munschy, C., Bely, N., Heas-Moisan, K., Pollono, C. and Jaquemet, S. (2021) Legacy and emerging organic contaminants in two sympatric shark species from Reunion Island (Southwest Indian Ocean): levels, profiles and maternal transfer. *Science of the Total Environment* 751, 10 January.

D'Mello, J.P.F. (2003) *Food Safety: Contaminants and Toxins.* CAB International, Wallingford, UK.

D'Mello, J.P.F. (2020) *Introduction to Environmental Toxicology.* CAB International, Wallingford, UK.

Donley, N. (2019) The USA lags behind other agricultural nations in banning harmful pesticides. *Environmental Health* 18, 44. https://doi.org/10.1186/s12940-019-0488-0

Frye, R.E., Cakir, J., Rose, S., Palmer, R.F., Curtin, P. and Arora, M. (2021) Prenatal air pollution influences neurodevelopment and behaviour in autism spectrum disorder by modulating mitochondrial physiology. *Molecular Psychiatry* 26, 1561–1577.

Fuertes, E., van der Plaat, D.A. and Minelli, C. (2020) Antioxidant genes and susceptibility to air pollution for respiratory and cardiovascular health. *Free Radical Biology and Medicine* 151, 88–98.

Grey Monroe, J., Srikant, T., Carbonell-Bejerano, P., Becker, C., Lensink, M. *et al.* (2022) Mutation bias reflects natural selection in *Arabidopsis thaliana. Nature* 2022. https://doi.org/10.1038/s41586-021-04269-6

Haycox, S. (2020) Crude oil pollution III. *Exxon Valdez* contamination: ecological recovery, a case study. In: D'Mello, J.P.F. (ed.) *A Handbook of Environmental Toxicology: Human Disorders and Ecotoxicology.* CAB International, Wallingford, UK, pp. 320–333.

Heneberg, P., Bogusch, P., Astapenková, A. and Řezáč, M. (2020) Neonicotinoid insecticides hinder the pupation and metamorphosis into adults in a crabronid wasp. *Scientific Reports* 10, 7077. https://doi.org/10.1038/s41598-020-63958-w

Johnson, N.M., Hoffmann, A.R., Behlen, J.C., Lau, C., Pendleton, D. *et al.* (2021) Air pollution and children's health: a review of adverse effects associated with pre-natal exposure from fine to ultrafine particulate matter. *Environmental Health and Preventive Medicine* 26, 72. https://doi.org/10.1186/s12199-021-00995-5

Katner, A.L. and Mielke, H.W. (2020) Lead poisoning. In: D'Mello, J.P.F. (ed.) *A Handbook of Environmental Toxicology: Human Disorders and Ecotoxicology.* CAB International, Wallingford, UK, pp. 371–383.

Laneve, G., Bruno, M., Mukherjee, A., Messineo, V., Giuseppetti, R. *et al.* (2022) Remote sensing detection of algal blooms in a lake impacted by petroleum hydrocarbons. *Remote Sensing* 14, 121. https://doi.org/10.3390/rs14010121

Lee, S., Ko, E., Lee, H. and Shin, S. (2021) Mixed exposure of persistent organic pollutants alters oxidative stress markers and mitochondrial function in the tail of zebrafish depending on sex. *International Journal of Environmental Research and Public Health* 18, 9539. https://doi.org/10.3390/ijerph18189539

Lin, C.-Y., Lee, H.-L., Hwang, Y.-T., Huang, P.-C., Wang, C. *et al.* (2020) Urinary heavy metals, DNA methylation and subclinical atherosclerosis. *Ecotoxicology and Environmental Safety* 204: 111039. https://doi.org/10.1016/j.ecoenv.2020.111039

Megson, D., Brown, T., Rhys Jones, G., Robson, M., Johnson, G.W. *et al.* (2022) Polychlorinated biphenyl (PCB) concentrations and profiles in marine mammals from the North Atlantic Ocean. *Chemosphere* 288: 132639. https://doi.org/10.1016/j.chemosphere.2021.132639

Melbourne, C.A., Shrine, N., Chen, J. and Wain, L.V. (2022) Genome-wide gene-air pollution interaction analysis of lung function in 300,000 individuals. *Environmental International* 159: 107041. https://doi.org/10.1016/j.envint.2021.107041

Melloni, B. (2020) Radon I. lung cancer risks. In: D'Mello, J.P.F. (ed.) *A Handbook of Environmental Toxicology: Human Disorders and Ecotoxicology.* CAB International, Wallingford, UK, pp. 475–483.

Mudway, I.S., Kelly, F.J. and Holgate, S.T. (2020) Oxidative stress and air pollution. *Free Radical Biology and Medicine* 151, 2–6.

Owens, A.C.S., Cochard, P., Durrant, J., Farnworth, B., Perkin, E.K. and Seymoure, B. (2020) Light pollution is a driver of insect declines. *Biological Conservation* 241: 108259. https://doi.org/10.1016/j.biocon.2019.108259

Pavlov, V., de Aguiar, V.C.M., Hole, L.R. and Pongrácz, E. (2022) A 30-year probability map for oil spill trajectories in the Barents Sea to assess potential environmental and socioeconomic threats. *Resources* 11, 1. https://doi.org/10.3390/resources.11010001

Pfeffer, P.E., Donaldson, G.C., Mackay, A.J. and Wedzicha, J.A. (2019) Increased chronic obstructive pulmonary disease exacerbation of likely viral etiology follow elevated ambient nitrogen oxides. *American Journal of Respiratory and Critical Care Medicine* 199(5). https://doi.org/10.1164/rccm.201712-2506OC

Richardson, E.A., Pearce, J., Tunstall, H., Mitchell, R. and Shortt, N.K. (2013) Particulate air pollution and health inequalities: a Europe-wide ecological analysis. *International Journal of Health Geographics* 12, 34. https://doi.org/10.1186/1476-072X-12-34

Tian, Y., Xiang, X., Juan, J., Song, J., Cao, Y. *et al.* (2018) Short-term effects of ambient fine particulate matter pollution on hospital visits for chronic obstructive pulmonary disease in Beijing, China. *Environmental Health* 17, 21. https://doi.org/10.1186/s12940-018-0369-y

Vishlaghi, N. and Lisse, T.S. (2020) Exploring vitamin D signalling with skin cancer. *Clinical Endocrinology* 92, 273–281.

Vogel, C.F.A., Van Winkle, L.S., Esser, C. and Haarmann-Stemmaan, T. (2020) The aryl hydrocarbon receptor as a target of environmental stressors – implications for pollution mediated stress and inflammatory responses. *Redox Biology* 34: 101530. https://doi.org/10.1016/j.redox.2020.101530

Zhang, S.-J., Zeng, Y.-H. and Zhou, J. (2022) The structure and assembly mechanisms of plastisphere microbial community in natural marine environment. *Journal of Hazardous Materials* 421: 126780. https://doi.org/10.1016/j.jhazmat.2021.126780

Appendix

Table 1 provides answers and guidance for selected supplementary questions listed in the order of appearance in each chapter. For solutions to other questions, students are advised to consult D'Mello *et al.* (1991) and D'Mello (2003, 2020a, b). Updates by Combi *et al.* (2022), Melila *et al.* (2022) and Zoran *et al.* (2022) are also suggested.

Table 1. Answers and guidance for supplementary questions in this volume.

Chapter	Topic	Answer
1	Epidemiology	Statistical correlation of population responses to environmental factors such as pathogens or pollutants
	Sigmoid curve	S-shaped dose–response curve
	NOAEL	No observed adverse effect level
	Asthma	Respiratory illness triggered and/or exacerbated by allergens and pollutants
	Mitochondria	Critical role in energy metabolism of living organisms
	Endoplasmic reticulum	Site of protein synthesis
	Hydrophilic	Affinity for water
	Lipophilic	Affinity for lipids/fats
	Blubber	Subcutaneous fatty tissue in sea mammals
	Biopsies	Examination of tissue samples obtained from living animal/human
2	Photosynthesis	Light-induced synthesis of sugars from carbon dioxide and water in the chloroplasts of plants and algae

Continued

Table 1. Continued.

Chapter	Topic	Answer
	Major chemical constituents of fertilizers	Nitrogen, phosphorus and potassium
	Farmer's lung disease	Caused by chronic inhalation of fungal spores from stored grain, hay and straw
	Nephropathy	Kidney damage
	Cord blood	Blood in the umbilical cord and placenta
	Leguminous plants in the garden	Peas and beans
	Glycosides	Compounds with attached glucose molecule
	HCN	Hydrogen cyanide
	Cassava	Tropical starchy tuber used as food
3	Cardiovascular disease	Disorder of the heart and vascular system, for example, hypertension
	Neurodegenerative syndromes	Disorders of the nervous system, for example, Alzheimer's disease
	Insulin	Hormone involved in regulation of glucose status in the body
	Free radicals	Unstable reactive atoms and molecules; by-products of cellular metabolism
	Macrophages	Role in detection, phagocytosis and elimination of bacteria
	Sulfuric acid	H_2SO_4
	Heterogeneous	Not uniform
	Redox	Coupled reduction–oxidation reactions
4	Subcutaneous	Under the skin
	Biomagnification	Sequential concentration along the food chain
	Adduct	Direct addition of one molecule to another to yield a single product
	Biphenyls	Organic compounds containing two fused phenyl rings
	Follicle-stimulating hormone	Follicle stimulating hormone regulates gametogenesis and synthesis of sex steroid hormones in mammals. Luteinizing hormone promotes ovulation in females and production of androgen in males.

Continued

Table 1. Continued.

Chapter	Topic	Answer
	Vitamin A	Retinol, an essential nutrient for humans (and other animals)
	Sodium channels	Formed by proteins in the plasma membrane for transport of sodium
	Dopamine	Neurotransmitter in the brain
	Muscarinic	Signalling network in the central and peripheral nervous system
	Systemic fungicide	Protectant acting within plants
	Genotoxicity	DNA damage
	Surfactant	Additive used to improve surface contact with active agent
	Symbionts	Microbes acting in mutual relationships with other organisms
	Circular DNA	Closed-loop DNA occurring in plasmids and mobile genetic elements
	Oestrogen	Female reproductive hormone
	Silviculture	Forestry
5	'Cancer alley' toxins	volatile organic compounds (VOCs)
	Dispersant used in oil spills	Corexit
	Organic components of fracking brines	Benzene, toluene, ethylbenzene and xylene
	Titania suboxides	Nanoparticles arising in coal combustion
6	Essential mineral element	Iron, among several others
	Flint TV movie	Available at: https://www.imdb.com/title/tt6397426 (accessed 7 March 2022)
	Organic mercury	Methylmercury
	Cysteine	Sulfur-containing amino acid
	Smelting	Processing of mineral ores
	Cognition	Understanding/reasoning
	Tight junctions	Membrane barriers at distinct points where two adjacent cells join together
	Cell-mediated immune response	Local immune reaction in tissues and organs
	Teratogen	Agent causing developmental abnormalities

Continued

Table 1. Continued.

Chapter	Topic	Answer
	Creutzfeldt-Jacob disease	Brain disorder
	Plaque	Fatty or other deposits in blood vessels or brain tissue
	Bradykinesia	Impaired voluntary movements
	Chromium	Cr
	Hexavalent	Valency of six – a specific oxidation state
7	'Great Pacific Garbage Patch'	For map see D'Mello (2020b)
	Pharmaceuticals	Categories include antibiotics, chemical contraceptives and illegal drugs
	M cells	Specialized epithelial cells in intestinal Peyer's patches
	Monomers	Constituents of complex polymers
	Oestradiol	Oestrogenic activity
	Bacteriophages	Viruses infecting bacteria
	Vibrio cholerae	Causal pathogen of cholera
	Disease transmitted by insect vector	Malaria
8	Tritium	Radioactive isotope of hydrogen
	Thyroid gland	Source of thyroxine
	Radioisotope decay	Breakdown
	miRNA	MicroRNA
	Vitamin D precursor in skin	7-Dehydrocholesterol
9	Horizontal gene transfer	Direct exchange or transfer of genetic material between populations of closely related bacteria or other species
	Signalling compounds	Endogenous agents such as salicylic acid and jasmonic acid that sense and instigate synthesis of defence molecules for protection of host plants from biotic and abiotic stressors
	Anaerobic	Absence of air/oxygen
	Dehalogenation	Removal of halogens such as chlorine and bromine
	Commensal microbes	Organisms residing in organs, particularly in the gut
	Hypoxia	Low oxygen status, for example in surface waters

Continued

Table 1. Continued.

Chapter	Topic	Answer
	Anoxia	Absence of oxygen, for example in surface waters
10	Carbon neutrality	Carbon emissions offset by mitigating actions
	Cardiac arrest	Heart attack
	Pesticides in aerosols	DDT, organophosphate insecticides and glyphosate, among several other pesticides
	Superfund site	Remediation site in the event of major disasters, for example the *Deepwater Horizon* accident (USA only)
	Cetaceans	Whales and dolphins
	Tuberculosis	Lung disease
	Ischaemic stroke	Caused by blood clot in brain
	Rhinoviruses	Viruses causing common cold
	Insulin resistance	Reduced response to endogenous insulin
	Carcinogenesis	Stages: initiation, promotion and progression
	Trimester	One of three phases in pregnancy
	Attention deficit hyperactive disorder	Behavioural condition, particularly in children
	2,4-dichlorophenoxyacetic acid	Short form: 2,4-D
	Ligand	Molecule that binds to another, often larger, molecule
	Transcription factors	Proteins that bind to DNA to influence the rate of gene transcription to mRNA
	Microbiota	Microorganisms
	Osmoregulation	Regulation of osmotic pressure in body fluids by controlling water and electrolyte balance
	Pelagic and demersal fish	Pelagic are coastal and oceanic fish at moderate depths of sea; demersal are deep-water species
	Soil fumigants	Used against soil-borne pests
	Calcium in human nutrition	Essential nutrient; major component of bones and teeth

References

Combi, T., Montone, R.C., Corada-Fernándex, C., Lara-Martín, P.A., Gusmao, J.B. *et al.* (2022) Persistent organic pollutants and contaminants of emerging concern in spinner dolphins (*Stenella longirostris*) from the Western Atlantic Ocean. *Marine Pollution Bulletin* 174: 113263. https://doi.org/10.1016/j.marpollbul.2021.113263

D'Mello, J.P.F. (2003) *Food Safety: Contaminants and Toxins.* CAB International, Wallingford, UK.

D'Mello, J.P.F. (2020a) *A Handbook of Environmental Toxicology: Human Disorders and Ecotoxicology.* CAB International, Wallingford, UK.

D'Mello, J.P.F. (2020b) *Introduction to Environmental Toxicology.* CAB International, Wallingford, UK.

D'Mello, J.P.F., Duffus, C.M. and Duffus, J.H. (1991) *Toxic Substances in Crop Plants.* The Royal Society of Chemistry, Cambridge.

Melila, M., Rajaram, R., Ganeshkumar, A., Kpemissi, M., Pakoussi, T. *et al.* (2022) Assessment of renal and hepatic dysfunction by co-exposure to toxic metals (Cd, Pb) and fluoride in people living near an industrial zone. *Journal of Trace Elements in Medicine and Biology* 69: 126890. https://doi.org/10.1016/j.jtemb.2021.126890

Zoran, M.A., Savastru, R.S., Savastru, D.M., Tautan, M.N., Baschir, L.A. *et al.* (2022) Assessing the impact of air pollution and climate seasonality on COVID-19 multiwaves in Madrid, Spain. *Environmental Research* 203: 111849. https://doi.org/10.1016/j.envres.2021.11189

Index